Water Technology for Developed and Developing Countries

Topical Conference at the 2011 AIChE Annual Meeting

Minneapolis, Minnesota, USA
16-21 October 2011

ISBN: 978-1-61839-576-4

Printed from e-media with permission by:

Curran Associates, Inc.
57 Morehouse Lane
Red Hook, NY 12571

Printed by Curran Associates, Inc. (2012)

For permission requests, please contact AIChE
at the address below.

AIChE
3 Park Avenue
New York, NY 10016-5991

Phone: (203) 702-7660
Fax: (203) 775-5177

www.aiche.org

Additional copies of this publication are available from:

Curran Associates, Inc.
57 Morehouse Lane
Red Hook, NY 12571 USA
Phone: 845-758-0400
Fax: 845-758-2634
Email: curran@proceedings.com
Web: www.proceedings.com

TABLE OF CONTENTS

(115a) 230097 A Global Challenge to Adsorption Processes in Public Water Supply Production: Mechanisms, Effects, Design Considerations, and Strategic Aversion 1
Walter J. Weber Jr.

(115b) 233204 Water Challenges and Opportunities In the Emerging World 30
Adil Dhalla

(115c) 233582 Epa's Research and Development Program for Innovative Water Technologies 31
Thomas Speth

(115d) 238104 Water Treatment Opportunities Utilizing the Dramatic Nano-Fluidic Properties of Carbon Nanotube Membranes 32
Bruce J. Hinds

(185a) 240280 A Perspective on the Impact of Material Sciences Innovations and Sustainable Development Considerations on Future Membrane Developments In the Water Industry 33
Hervé Buisson

(185b) 240510 ZW500 MBR: 15+ Years of Evolution 34
Geert-Henk Koops

(185c) 226900 Gas Sparging for Membrane Fouling Control 35
Pierre P. Brub

(185d) 240294 Fouling Causes and Control Strategies for Membrane Bioreactors In Municipal Wastewater Treatment 36
Hongde Zhou

(185e) 240781 Ozonation of a Non-Nitrified Secondary Effluent to Reduce Microfiltration Fouling 37
R. Shane Trussell

(185f) 240512 Future Directions for Wastewater Treatment 38
Pierre Côté

(270a) 233179 Nanomaterials and Recent Developments In Membrane Technology In the Middle East 39
Suzana Pereira Nunes

(270b) 231084 Trend of Membrane Technology for Water Treatment In Japan 40
Hideto Matsuyama

(270c) 237704 Water Treatment R&D In Korea 41
Seockheon Lee

(270d) 231213 Applications of Membrane Technologies for Water Treatment In Taiwan 42
Da-Ming Wang, Kuo-Lun Tung, Juin-Yih Lai

(270e) 220665 Green Preparation of PVDF Microfiltration Membranes Via Thermally Indueced Phase Separation Method Using Diphenyl Carbonate and Diphenyl Ketone As Diluents 43
Xiao-Lin Wang

(270f) 222667 Water Sustainability and R & D In Singapore 45
Tai-Shung Chung

(301a) 236518 Opportunities for Government to Influence Technology Innovation In the Water Industry 46
Jeffrey Q. Adams

(301b) 235351 Navy Shipboard Desalination 49
Ian Peek, David Nordham, Paul Armistead

(301c) 228858 Regulatory Compliance and Water Treatment Advances: From Research to Application 52
John Albert

(301d) 221556 Inland Brackish Groundwater As a Resource: Federal Research, Issues, and Challenges 53
Michelle Chapman

(301e) 223180 Electrodialysis Reversal Pilot Studies Performed At the Brackish Ground Water National Desalination Research Facility (BGNDRF) In Alamogordo, NM 56
Neil E. Moe, John Barber, Ali Sharbat, Abbas Ghassemi

(301f) 223179 Feasibility Study of High Cut Rate of Dissolved Mineral Salts In Electrodialysis Reversal (EDR) Process 57
Ali Sharbat, Neil E. Moe, John Barber, Abbas Ghassemi

(399a) 240259 Treatment of Ammunition Wastewater Using Ultrasound-Fenton Process (Invited Keynote Speaker) 61
Yangang Li, Wenpin Hsieh, Rovshan Mahmudov, C. P. Huang
(399b) 239875 COD Reduction From Leachate with An Advanced Oxidation Process (AOP) 77
Kashinath Banerjee, Tapas Das
(399c) 232330 Photocatalytic Decomposition of Geosmin and 2-Methylisoborneol by M-InVO4/TiO2 with Visible Light In Recirculating Aquaculture Systems 78
Sandra L. Pettit, Louis J. Rufo, John T. Wolan
(399d) 237733 Advanced Oxidation for Water Processing In Microstructured Reactors 79
Kevin Caple, Nick Kusanto, Kenneth Williamson, Jeffrey A. Nason, Goran N. Jovanovic, Alexandre F.T. Yokochi
(399e) 236878 Kinetics of Oxidation of Dyes In Effluents Using a Porous Electrode Ozonator 80
Alexander P. Mathews, Kishora K. Panda
(399f) 237027 Adsorptive-Reactive Nanoparticulate Systems for the In Situ Remediation of Chlorinated Hydrocarbons 90
Bhanukiran Sunkara, Jingjing Zhan, Yingqing Wang, Vijay T. John, Jibao He, Gary McPherson
(465a) 218175 A Different Look At Water: Water for Cities and the Water-Energy Nexus 91
Snehal Desai
(465b) 230251 Reducing Energy Consumption In Reverse Osmosis Desalination: Cyclic or Multi-Stage Operation? 92
Aihua (Richard) Zhu, Panagiotis D. Christofides, Yoram Cohen
(465c) 223478 Hierarchical Nanoporous Filtration Membranes Enabled by a Roll-to-Roll Compatible Bottom-up Technology 94
Peng Jiang
(465d) 234261 Fundamental Water and Ion Transport Properties of Sulfonated Polysulfone 95
Joseph Cook, Chang Hyun Lee, Wei Xie, Benny D. Freeman, James E. McGrath
(465e) 231354 Preparation of Robust Hybrid Silica Membranes for Reverse Osmosis 96
Toshinori Tsuru, Rong Xu, Jinhui Wang, Masakoto Kanezashi, Tomohisa Yoshioka
(465g) 230386 Investigation of Protein Adsorption Behavior On Polymer Surfaces Using Atomic Force Microscopy and Quartz Crystal Microbalance with Dissipation 97
Toru Ishigami, Keisuke Hirami, Masatoshi Hashino, Noboru Kubota, Yoshikage Ohmukai, Eiji Kamio, Tatsuo Maruyama, Hideto Matsuyama
(545a) 237788 Improving Electrospun Nanofiber Strength for Filtration Applications 100
Liwei Huang, Ngoc Bui, Jason Arena, Seetha Manickam, Jeffrey R. McCutcheon
(545b) 231622 Surface Modification of Membranes to Improve Fouling Resistance 101
Sirirat Kasemset, Albert Lee, Daniel J. Miller, Benny D. Freeman, Mukul M. Sharma
(545c) 237546 Naturally-Functionalized Antimicrobial Sand (f-sand) Using Moringa Oleifera Seeds 102
Darrell Velegol, Huda A. Jerri, Kristin J. Adolfsen, Lauren R. McCullough, Stephanie B. Velegol
(545e) 222190 Nano-Pipette Directed Motion of Functionalized Transmembrane Channel 103
Meenakshi Dutt, Olga Kuksenok, Anna C. Balazs
(575a) 232305 Treatment of Produced Water by Ceramic Membranes 104
Bruce A. Bishop, Stanton Smith, Lnsp Nagghappan
(575b) 225346 Electrosteric Stabilization and Covalent Functionalization of Iron Nanoparticles In a Mixed Matrix Membrane 105
Lauren F. Greenlee, Jessica D. Torrey
(575c) 234906 TiO2-Enhanced Polymer-Derived Ceramic Composite Membranes for Water Purification 106
Jessica D. Torrey, Lauren F. Greenlee
(575d) 238781 Silica-Polymer Mixed Matrix Membranes for Mercury and Silver Sorption 107
Noah D. Meeks, Dibakar Bhattacharyya
(594a) 220791 Characterization of Pecan Shell Activated Carbon In Aqueous Chromium Systems 108
Venkata Daram
(594b) 236337 Biosorption of Lead by Spent Coffee Ground: Kinetic and Isotherm Studies 109
N. E. Davila-Guzman, F. J. Cerino-Córdova, J.R. Rangel-Mendez, P.E. Diaz-Flores
(594c) 232853 Development of Waste Water Treatment Process with Sorbents Derived From Waste Concrete Sludge 118
Akihiro Yamasaki, Takeshi Sasaki, Yusuke Tsunashima, Atsushi Iizuka, Yuka Sakai, Teruhisa Hongo, Yukio Yanagisawa
(594d) 215763 Adsorption of Phenol From Aqueous Solutions Using Activated Carbon Prepared From Crofton Weed 119
Chao Hua, Runhu Zhang, Fang Bai
(594e) 227264 Metals Adsorption From Bioleaching Solution of Printed Circuit Board 120
Len Foong Koong, Gordon Mckay, John Barford, Koon Fung Lam

(658a) 222410 Ultrafiltration/Nanofiltration Based Technique for the Removal of Endocrine Disruptors From Surface Water 121
Ana Gil, Georgina Sarmento, Miguel Minhalma, Elizabeth Duarte, Maria Norberta De Pinho
(658b) 240601 The Importance of a Well-Designed Cross-Flow System In the Removal Study of Adsorbing Estrogens by Nanofiltration Membranes 126
Andrea JC Semiao
(658c) 240560 SW-RO Membranes for the Removal of Ammonium Thiocyanate From Coking Wastewaters In Steel Manufacturing 128
Susana Luque, Jose R. Álvarez, M. El-Mansour, F.E. Antón, B. González, B. P. Vivas
(658d) 238755 Environmental and Economic Evaluation of Wastewater and Water Disinfection Technologies 129
Tapas Das
(658e) 225940 Chemical Properties and Degradation Efficiency of Radio-Frequency Plasmas In Water 130
Selma Mededovic
(658f) 235177 Determination of Bisphenols Using Ultraviolet Spectroscopy 131
Alfred R. Conklin Jr., Jessica Veite, Jennifer Wilson
(659a) 240146 Static Mixing Spacers for Spiral Wound Modules 137
Ashkan Iranshahi, Glenn Lipscomb
(659b) 238261 Atomistic Simulation of Water and Salt Transport In RO Membrane Active Layers 138
Jed W. Pitera, Young-hye Na, Ankit Vora, Geraud Dubois
(659c) 229037 Dynamic Data-Driven Models of Reverse Osmosis Desalination Systems 139
Xavier Pascual, Han Gu, Alex Bartman, Aihua (Richard) Zhu, Anditya Rahardianto, Jaume Giralt, Robert Rallo, Joan Ferrer-Giné, Francesc Giralt, Yoram Cohen
(659d) 225697 The Role of External Electric Fields In Enhancing Water Treatment Technologies 140
Sohail Murad
(659e) 234297 Effect of Convective Hindrance On Microfiltration of Rod Shaped Particles 141
Ruth E. Baltus, Basavaraju Nagendra Agasanapura, Shankar Chellam
(691a) 240857 Fouling and Cleaning Behavior of Forward Osmosis Membrane (keynote) 143
Baoxia Mi
(691b) 239048 Permeation of Interacting Solutes In Forward Osmosis 144
William Phillip, Jui Shan Yong, Menachem Elimelech
(691c) 223502 Network-Based Finite Element Analysis for Studying the Effect of Support Structure On Internal Concentration Polarization During Forward Osmosis 145
Weiyi Li, Yiben Gao, Chuyang Y. Tang
(691d) 237858 New Approaches to Forward Osmosis Membrane Design 146
Ngoc Bui, Jason Arena, Jeffrey R. McCutcheon
(691e) 233036 Development of a Pilot-Scale Forward-Osmosis Desalination Process 147
Young Kim, Dong-Wook Oh, Yu-Chang Kim, Kong Hoon Lee, Jungho Lee
(691f) 237618 Zero-Energy Osmotic Water Purifier for Ethiopia: Providing Clean Drink to Vulnerable Populations 148
Ethan Butler, Jeffrey R. McCutcheon
(762a) 232021 Desalination of the Colorado River Water: A Hybrid Approach 150
Anand Venkatesan, Phillip C. Wankat
(762b) 225331 Toward Sustainability and Low Cost In Electro-Dialysis Reversal Desalination 151
Maung Thein Myint, Abbas Ghassemi
(762c) 227617 Municipal Wastewater Treatment Using Electrocoagulation and Adsorption for Reuse In Industrial Processes and Green Areas 158
Hector Moreno, Alejandro Romero, Jewel Andrew Gomes, J. R. Parga, Eric Peterson, Cristina Garcia, Pantaleona Campa
(762d) 237816 Transport and Remediation of Chromium (VI) In Groundwater Using An Integrated Electrokinetic and Nanoscale Particle Technology 171
Charlotte Chan, Ryan Thacher, Lewis Hsu, Massoud Pirbazari
(762e) 236367 Starvation of a Mixed Culture of Chlorella Vulgaris and Scenedesmus Sp. for Enhanced Removal of Ammonia From Wastewater Effluent and Its Impact On Lipids Formation 173
Bala Padmini Lingaraju, Jinsoo Kim, Joo-Youp Lee, Kaniz F. Siddiqui, Zhouyang Liu
(762f) 237676 Influent Solids Retained High Biomass Reactor Treating Dairy Manure Wastewater and Kinetics Study 175
Jingwei Ma, Liang Yu, Shulin Chen
Author Index

A Global Challenge to Adsorption Processes in Public Water Supply Production: Mechanisms, Effects, Design Considerations, and Strategic Aversion.

Walter J. Weber, Jr., The Gordon M. Fair and Earnest Boyce Distinguished University Professor, Emeritus, Department of Chemical Engineering, The University of Michigan, Ann Arbor, MI 48109-2125

Abstract

Fixed-bed reactors (FBRs) containing granular microporous activated carbons or polymeric resins comprise integral parts of many public potable water supply treatment systems. These adsorbents and reactor configurations are highly effective in removing potentially harmful synthetic organic contaminants found increasingly in raw source waters. The existence in most raw source waters of complex and highly variable natural organic materials (NOMs) is essentially a concomitant reality, While NOMs themselves pose no direct contamination problems, various molecular fractions thereof are in fact also actively adsorbed, but in preloading" mode" by activated carbons and synthetic polymer adsorbents. This adsorbent "preloading" by NOMs has profound negative effects on the subsequent ability of the adsorbents to remove the potentially hazardous organic contaminants for which they were intended. Unfortunately, these effects are seldom factored adequately into FBR initial designs. This paper presents a rational approach by which the impacts of NOM preloading can be incorporated into the design of FBR adsorber systems on a source-water specific basis. As suggested by its title, the scope of the paper is broad. It begins with an examination of phenomena and mechanisms associated with NOM preloading of FBR adsorbers. The results of a series of investigations from which we have evolved a quantitative approach to the integration of preloading effects in the design of FBR adsorbers are then detailed and a specific case study described. By way of an enticing brief conclusion, a clever means for Strategic Aversion of the global NOM challenge described is referenced.

Introduction

The principal purpose of adsorption systems in water treatment operations is to remove specific types of undesirable organic compounds such as halogenated solvents, pesticides, and fuel residues that pose threats to human health). Activated carbons and several synthetic organic polymers serve this function by providing large surface areas populated by different types of functional groups generated in various "activation processes".. Nuisance and potentially harmful organic compounds are generally present at relatively small concentrations (i.e., ppb levels) in most potable water sources. Conversely, a heterogeneous array of natural organic compounds (e.g., humic and fulvic acids) are commonly present in the same source waters at much higher concentrations (i.e., ppm levels). While compounds comprising this heterogeneous natural organic matter (NOM) are not necessarily harmful to public health on their own accord, some of their constituents interact with halogen-based disinfectants to form potentially harmful "disinfection by-products" (e.g., chlorine and bromine substituted aliphatic and aromatic compounds).

By design, activated carbon is a highly effective adsorbent for removing a broad range of different types of undesirable organic compounds targeted for removal during water treatment (i.e., "target" compounds). Given these inherent properties, GAC is thus effective also in removing many compounds comprising NOM in sources waters. This is desirable from one perspective; i.e., GAC reduces the number of NOM components available to color water, cause odors or tastes, and function as precursors for formation of disinfection by-products. From a different perspective however it is undesirable; i.e., the diverse array of chemical moieties comprising NOM compete effectively with target compounds for the surfaces and reactive sites of activated carbons and polymeric adsorbents, thus interfering with and diminishing its adsorption of target compounds.

Preloading Phenomena

On a one-to-one "molecular" basis, NOM moieties are generally not adsorbed as effectively as target contaminants. However, their presence in order-of-magnitude higher concentrations makes them effective competitors for adsorption sites on GAC. Another less obvious, and somewhat ironic, reason NOM moieties compete effectively for GAC sites is that most of them are in fact not as effectively adsorbed as most target contaminants. This behavior allows them more rapid access than target compounds to the fresh activated carbon regions of a typical fixed bed adsorber (FBR), as depicted by the wavefront behaviors shown in Figure 1. In accessing the fresh GAC more rapidly, the NOM moieties "preload" and "precondition" the fresh GAC in ways that potentially change its properties and reduce its effectiveness for adsorption of target contaminants that come along later. In other words, the fresh GAC intended for target compound removal rapidly becomes a "used" GAC of unknown, but generally lesser, effectiveness for subsequently removing target compounds.

Figure 1. Relative Wavefront Behaviors Leading to Preloading

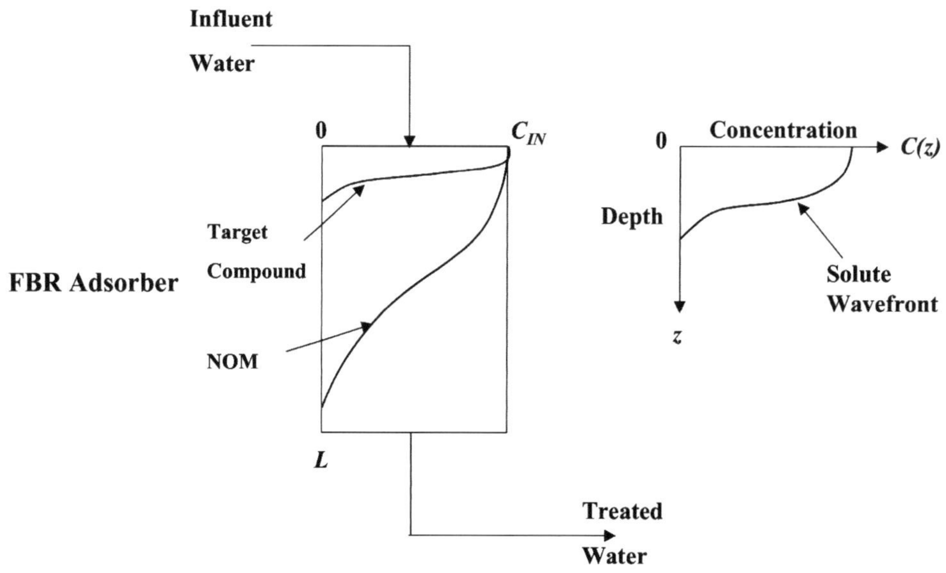

It would seem at first consideration a relatively simple, albeit costly, matter to add more carbon to account for the decreases in its capacity caused by adsorption of the NOM compounds. However, it is impossible to do this in a rigorous engineering manner without knowing the nature and magnitude of the effects of the NOM on the abilities and capacities of the GAC for target compound removal. Unfortunately, these effects can neither be predicted from any available adsorption theories nor can they be assessed by simple lumped-parameter NOM concentration measurements (e.g., total organic carbon). This is the case because: i) the water containing the NOM and the target compounds is markedly changed in character once it moves a short distance in the GAC bed; and ii) the character of the GAC once so contacted is changed as well from what it was initially. To make the situation even more complicated, the changes in each of these two media relate to the specific changes that occur in the other as functions of time and position in the

FBR. Such changes in turn relate to the chemical composition of the water being treated and to the flow characteristics of the FBR.

Adsorption of NOM

Figure 2 presents a schematic representation of the effects of various parameters on the adsorption of NOMs by GAC, plotted as functions of an important background water control variable; i.e., pH. This qualitative representation illustrates the variety of factors that can influence NOM adsorption behavior, which in turn determine how a given GAC will be affected by preloading.

Figure 2. Parameters Affecting NOM Sorption (Karanfil et al, 1998)

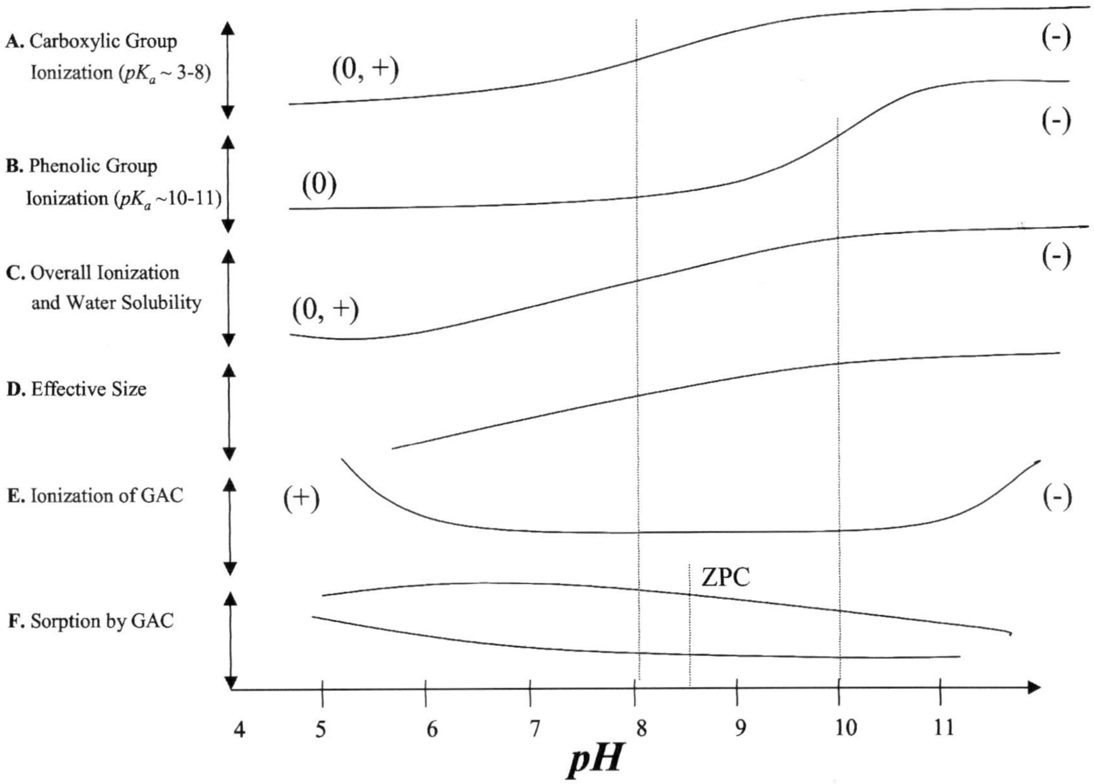

To provide a quantitative example of some of the effects shown in Figure 2, consider the equilibrium adsorption behaviors of the seven NOMs identified and characterized in

Table 1. To characterize equilibrium conditions for this example, we use the following modified Freundlich equation to represent the adsorption isotherm

$$q_e = K_{F,M}\,(C_e/D_o)^n \qquad (1)$$

where q_e is the solid phase concentration of NOM at equilibrium expressed as milligrams of total organic carbon (TOC) per gram of GAC, C_e is the corresponding solution phase concentration (mg TOC/L), D_o is the initial GAC dosage (g/L), $K_{F,M}$ is a Freundlich "affinity" parameter for the chemically heterogeneous NOM, and n is related to the magnitude of the adsorption driving force and to the distribution of energy sites on the GAC (Weber and DiGiano, 1996). The ratio C_e/D_o is the GAC-dosage normalized residual NOM concentration; i.e., the amount of unadsorbed NOM per unit of carbon mass present in the reactors used to collect individual q_e and C_e data points for isotherm plotting and fitting with Equation 1. This modification of the conventional Freundlich equation, which is identical to Equation (1) but does not normalize C_e for D_o, has been found suitable for characterizing the isotherms of heterodisperse polymers. Both the modified and conventional Freundlich equations have been used in different research projects described in this paper, the modified form given in Equation 1 has been used for the most part in the NOM adsorption equilibrium work described in the first half of this paper. The conventional form was employed in the case study example of preloading effects on target compound removal.

Table 1. NOM Characteristics (Karanfil et al, 1996)

NOM	Carbon/Oxygen Contents (%) C	O	O/C	Aromaticity (%)	Phenolic Acidity (meq/g TOC)	Carboxylic Acidity (meq/g TOC)	Weight-Average MW
Aldrich HA	50.7	31.4	0.47	57.7	3.6	7.9	4006
Soil HA (IHSS)	58.0	33.7	0.44	50.0	2.8	5.9	6202
Peat HA(IHSS)	56.6	35.4	0.47	47.0	3.5	7.0	5584
Leonardite HA (IHSS)	63.1	31.0	0.37	58.0	4.4	5.5	5867
Suwannee River FA (IHSS)	53.5	41.3	0.58	24.8	3.3	8.1	1920
Laurentian HA	51.9	39.9	0.58	33.4	5.7	8.9	3982
Laurentian FA	45.1	49.7	0.83	28.2	9.3	11.7	2402

Data obtained by Karanfil, *et al* (1996) for the individual isotherms for the seven NOMs identified in Table 1 are plotted in Figure 3, along with the best fit of the modified Freundlich model to each set of data. Values of $K_{F,M}$ and n for these different NOMs range from 6.6 to 1.7 (mg TOC/g GAC)$^{1-n}$ and from 0.43 to 0.22, respectively, the relative adsorption capacities of the GAC for the NOMs decline in the order: SRFA(1920)>AHA(4006)>LaHA(3982)>LaFa(2402)>LeHa(5867)~SHA(6202), where the quantities in parentheses are the number-average molecular weights of the NOMs determined by size exclusion chromatography.

Figure 3. Humic and Fulvic Acid Sorption Isotherms (Karanfil, 1996)

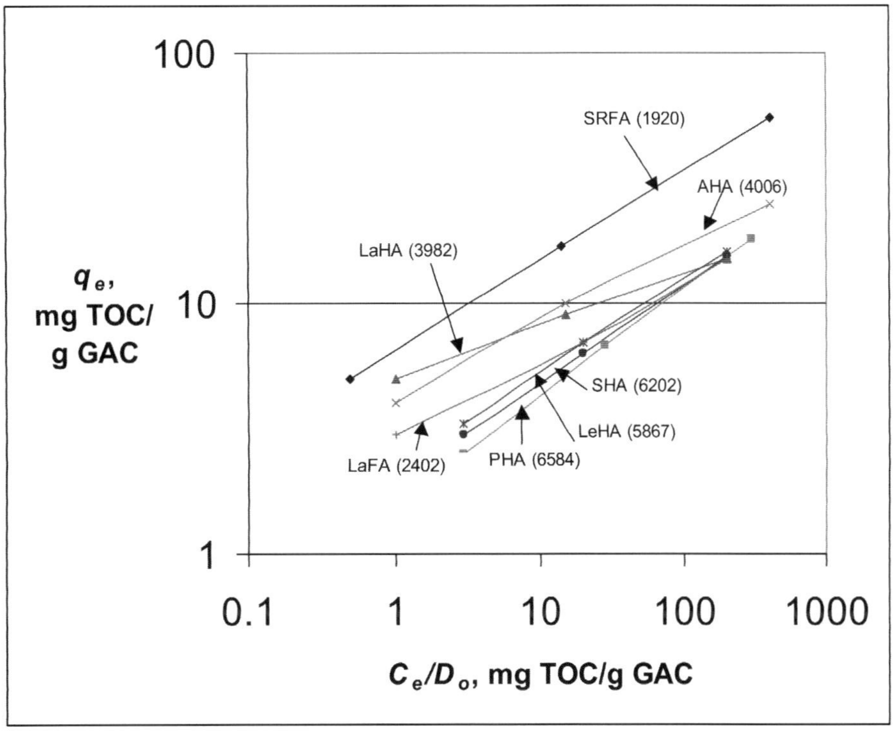

Figure 4 depicts the different adsorption behaviors of various size fractions of one of the NOMs identified in Table 1, i.e., the Laurentian Humic Acid (LaHA). Values of $K_{F,M}$ and n for the different size fractions of this LaHA were found to range from 18.9 to 3.5 (mg TOC/g GAC)$^{1-n}$ and from 0.12 to 0.25, respectively, compared to values of 5.3 (mg TOC/gGAC)$^{1-n}$ and 0.22 for the whole LaHA. It is clear from Figure 4 that, for the constituents of a given NOM mixture, the relative adsorption capacity of the GAC decreases with increasing molecular size. Figure 3, however, just as clearly indicates that other properties of NOM mixtures are also important determinants of adsorption, as suggested earlier in Figure 2.

Figure 4. Sorption Isotherms for Components of Laurentian HA

(Karanfil et al, 1996)

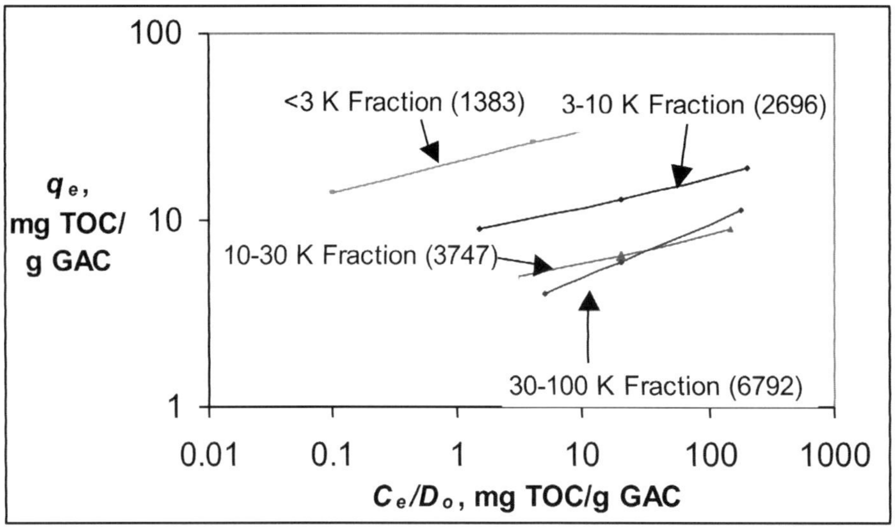

To bring the discussion relating to factors affecting NOM adsorption behavior to a close, Figures 5 and 6 after Karanfil *et al* (1998) show the relative effects of three general solution properties on the adsorption of the Laurentian Fulvic Acid (LaFA) identified in

Table 1. Parts A and B of Figure 5 show isotherms for adsorption of the LaFA from oxygenated and deoxygenated water for three different solution pH values and three different ionic strengths, respectively. Parts A and B of Figure 6 present corresponding parametric trends for the ratio $(K_{F,M})_{oxic}/(K_{F,M})_{anox}$ as functions of solution pH and ionic strength, respectively.

Figures 5. Laurentian HA Isotherms as Functions of pH (A) and Ionic Strength (B)

(Karanfil et al, 1998)

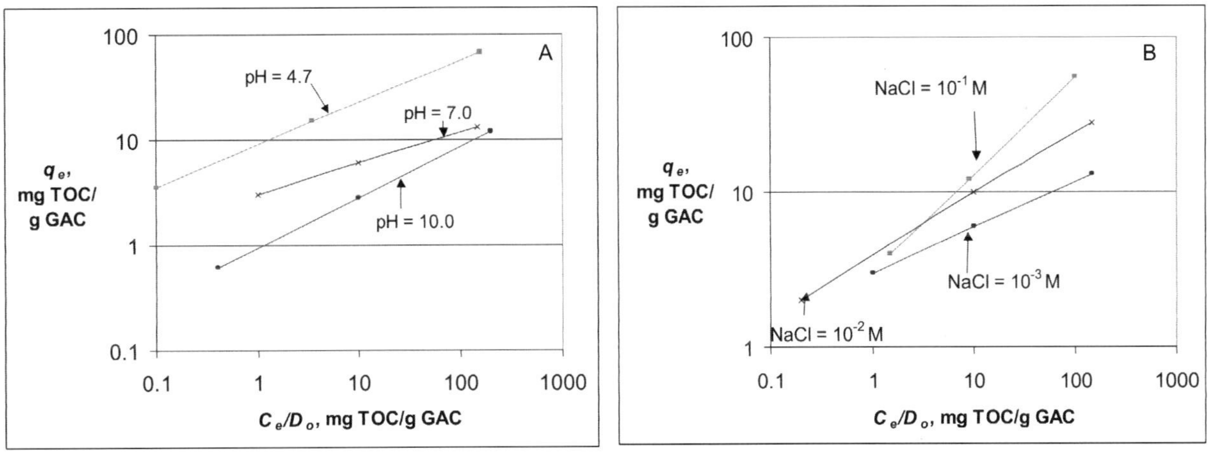

Figure 6. ***pH*** **(A) and Ionic Strength (B) Effects on Oxygen Sensitivity**

(Karanfil et al, 1998)

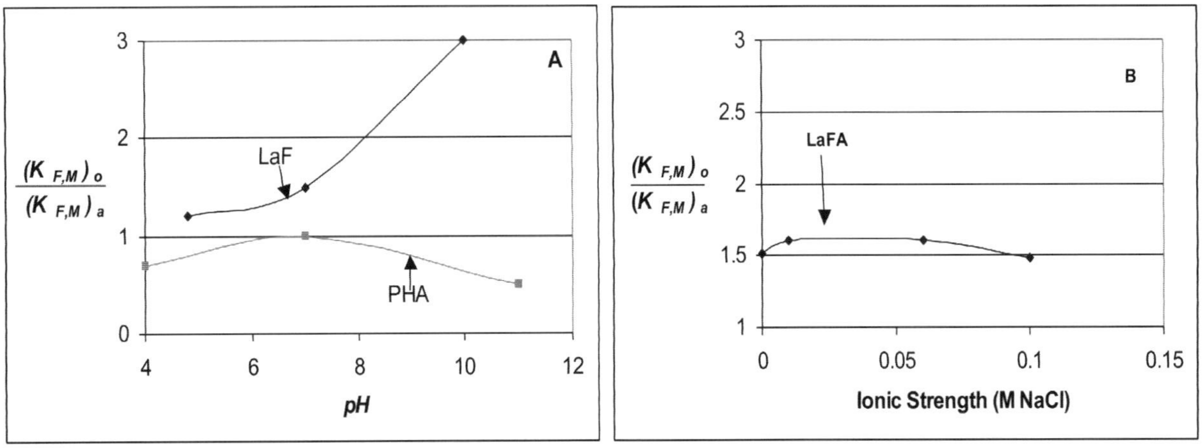

Preloading Mechanisms

The dual mechanism concept advanced by Carter *et al* (1992) involving sequential occupation of high energy surfaces and subsequent pore blocking as an explanation for preloading effects has more or less stood the test of time. This concept derives from observations of changes in the exponent n of the Freundlich isotherm model (modified or otherwise) as a function of increasing time of preloading of GAC in CMBRs and short-bed adsorbers (SBAs). As noted earlier in discussion of Equation 1, the parameter n is representative of the distribution of energy sites on solid adsorbents. The concept is depicted schematically in the series of four panels shown in Figure 7 with the n values given in the panels being actual experimental values measured for adsorption of NOM from Huron River water, a raw water source for the city of Ann Arbor, MI.

Figure 7. Schematic Representation of Preloading Mechanisms (Carter et al, 1992)

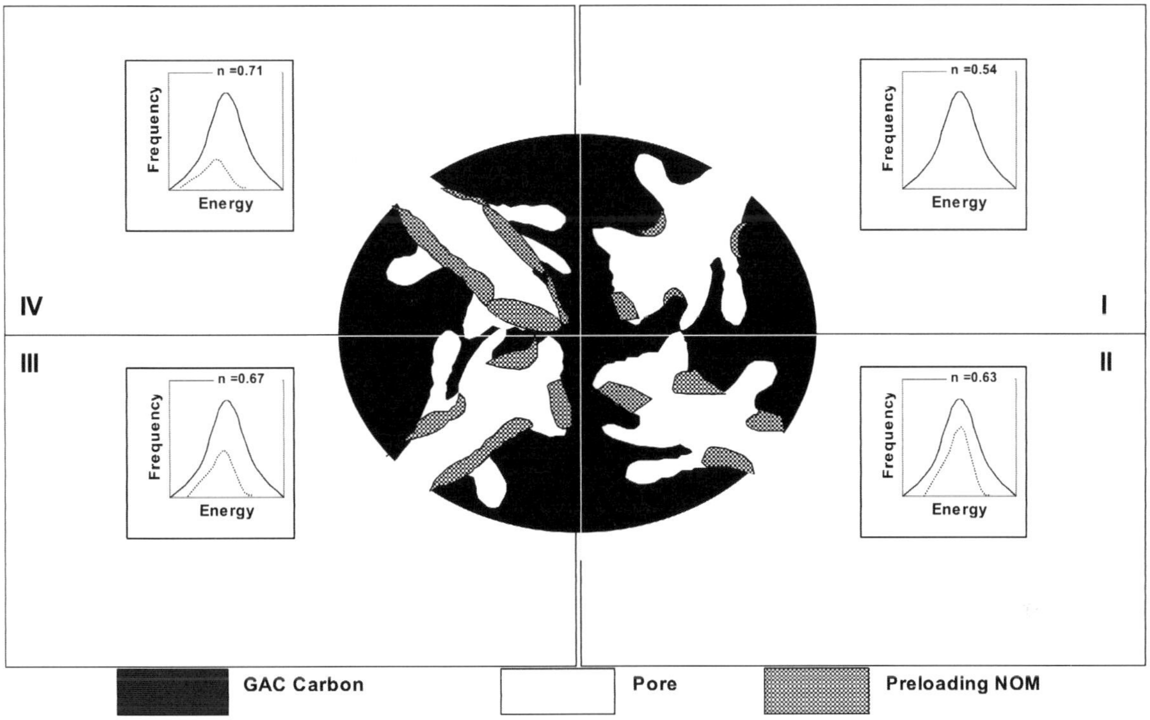

Panel I in Figure 7 represents an initial non-preloaded carbon reference state in which the surfaces and pores are free from NOM. In this state the GAC has highest energy site heterogeneity and lowest *n* value along with the corresponding hypothetical site energy distribution shown in Figure 7. Panel II depicts early preloading effects, with certain NOM species rapidly moving through the larger pores and occupying the most readily accessible high-energy sites on the GAC. The resulting site energy distribution is thereby narrowed from that of non-preloaded carbon, with losses occurring especially in high-energy regions. This is reflected in a rapid initial increase in *n* for early preloading. Panel III represents an intermediate stage in the preloading process as the smaller pore openings begin to be blocked and the remaining readily accessible high-energy sites are lost.

Losses in numbers of sites then begin to occur more evenly across all energy levels, resulting in a gradual leveling off of K_F (or $K_{F,M}$) and *n* values at extended preloading times. This plateauing or leveling of parameter values suggests that most of the readily accessible high energy sites have been exhausted, and that further decreases in capacity occur more slowly via ongoing pore blockage. Panel IV in Figure 7 represents the state of an extensively preloaded carbon. Primary capacity reduction at and beyond this state is mechanistically attributable principally to pore blockage, and all isotherms tend toward parallel slopes (i.e., equal n values).

The above mechanism is consistent with a distributed reactivity model (DRM) concept advanced to explain the observed sorption behaviors of heterogeneous soils and sediments in subsurface systems (Weber, *et al*, 1991). In the context of the present discussion, this model projects a sequential compression of an initially wide distribution of sites having different adsorption energies to a more narrowly distributed range by virtue of an initial and uncontested occupation of higher-energy sites on the GAC by selected NOM moieties, followed by a gradual shifting of the dominant capacity reduction mechanism to one of pore blockage by different NOM constituent species. The concept visualizes complex sorption phenomena as comprising one or more typically linear absorption processes, and one or more typically non-linear adsorption processes. Each absorption and adsorption contribution to the overall process manifests a different level of reactivity. For such situations, the sum of absorptive processes increases with increasing saturation of surfaces and pore volumes by organic matrices, and the overall sorption phenomenon thus trends to more linear overall behavior.

An initially fresh (non-preloaded) activated carbon accordingly may act to remove organic compounds primarily by highly non-linear adsorptive mechanisms that exhibit *n* values much lower than unity. This is shown in panel I of Figure 7. As the carbon is preloaded, however, NOM molecules begin to coat the surfaces of the pores, and perhaps even begin to fill some pore spaces completely, essentially forming an organic phase into which absorption now occurs. Panels II through IV of Figure 7 depict this progression of organic accumulation within the carbon pores. As the organic saturation of the carbon increases, the sorption balance shifts away from surface adsorption mechanisms and more toward absorption or "partitioning" mechanisms, with corresponding increases in *n* values. At infinite preloading time, all pore volumes would presumably be filled with NOM organic matrix, and the removal of the target compounds would occur primarily by a nearly linear absorption-like process.

Removal Effects on Target Compounds: A Case Study

As noted earlier and depicted in Figure 1, the most common reactor configuration for GAC applications in water and wastewater treatment is that of a fixed-bed reactor (FBR).

To illustrate the effects of preloading on FBR adsorption behavior with respect to target organic compounds we will examine a specific case study based on work by Carter and Weber (1994) involving a water treatment grade carbon (Filtrasorb 400, Calgon Corp., Pittsburgh, PA), a common chlorinated solvent target contaminant (trichloroethylene, TCE) and a public surface water supply source (Huron River water,

Ann Arbor, MI) having a background NOM as any other might be reasonably representative.

FBR Model

A pore diffusion model was used for calibration of rate data and simulation of all FBR breakthrough data. The model equations and solutions are based upon earlier work by Crittenden *et al.* (1986). The model was reduced to a set of simultaneous ordinary differential equations (ODEs) in time using the method of orthogonal collocation. The coefficients of the ODEs incorporate the isotherm and rate information provided by the equilibrium parameters K_F (note use of conventional Freundlich model form), n and the rate parameters kf (external film mass transfer coefficient), and Dp (internal pore diffusion coefficient). The breakthrough simulation was then obtained by calling a general ODE solver for stiff equations (LSODE) to solve the ODEs through a series of time steps until the total run time was met. The validity of this approach and the ability of the pore diffusion model to describe adsorption in fixed beds have been widely demonstrated. The conventional constant parameter version of the pore diffusion model was used for calibration of SBA rate data and for simulation of DBA adsorber data. A second version of the pore diffusion model referred to here as the time-variable parameter model was developed in which each of the isotherm and rate parameters was input as a function of model run time, e.g., $K_F = \phi(t)$. The isotherm and rate parameters were then updated at each time step, resulting in ODE coefficients that changed over the model run time according to the parametric input expressions.

GAC Preloading

The general chemical characteristics of the Huron River water (HRW) used for preloading are given in Table 2. This water was collected in one batch, dosed to a level of 100 mg/L with sodium azide, and stored under refrigerated conditions. Immediately prior to its use in both isotherm experiments and column studies the HRW was filtered through a 1.0-μm filter. A GAC of US Standard Sieve Size 30/40 (0.512mm arithmetic mean diameter) was preloaded with HRW in FBR adsorbers at a nominal surface loading of approximately 11 cm/sec. The four parallel beds of GAC were preloaded for one week, for two weeks, for three weeks, and four weeks, respectively. Breakthrough of TOC was measured as a function of time in each preloading bed. The mass of background NOM loaded onto the four-week bed was 27 mg TOC/g GAC.

Parameter	Value
Total Organic Carbon (mg/L)	7.4
pH	8.3
Alkalinity (mg/L as CaO_3)	248
Calcium (mg/L as CaO_3)	178
Hardness (mg/L as CaO_3)	260
Ionic Strength (mM as NaCl)	6.9
Mol wt of Organic Matter (amu) (peak weight)	1000-5000 (1860)

Table 2. Huron River Water (Carter and Weber, 1994)

Equilibrium Studies

Conventional Freundlich Isotherm constants for adsorption of TCE from HRW in CMBR systems for the cases of no preloading and preloading for two, three, and four weeks are presented in Table 3. The data indicate significant loss of equilibrium capacity with increased preloading time with the biggest drop occurring by the end of the second week. K_F values dropped from 1.39 to 0.287 over four weeks of preloading. Correspondingly, Freundlich *n* values increased from 0.542 to 0.713.

Table 3. Isotherm and Rate Parameters (Carter and Weber, 1994)

Preloading Time	K_F (1)	n (2)	k_f (3)	τ (2)
None	1.4	0.54	3.5	1.2
One Week	-	-	-	-
Two Weeks	0.49	0.63	1.4	5.0
Three Weeks	0.41	0.67	1.2	8.0
Four Weeks	0.29	0.71	1.0	6.0

(1) Based on $q_e = K_F C_e^n$ (with q_e = (μg/mg) and C_e = (μg/L))
(2) Dimensionless
(3) cm/sec * 10^3

Rate Studies

TCE rate parameters were gathered in the background waters for various levels of preloading using the short-bed adsorber (SBA) technique developed by Weber and Liu (1980). In the pore diffusion model described above the relevant rate parameters are k_f, the film transfer coefficient which characterize mass-transfer resistance across a hydrodynamic boundary layer surrounding a carbon particle, and D_p, the intraparticle pore diffusion coefficient which quantifies TCE diffusion through the liquid contained within internal particle pores. The SBA is designed to be sufficiently short so that virtually immediate solute breakthrough occurs. Thus, for early time data, (i.e., before significant intraparticle mass transport occurs) simulations using the pore diffusion model are sensitive only to changes in the film transfer coefficient, k_f. By calibrating only early time data, an explicit determination of k_f is made. Once k_f is determined, it is held constant and the entire SBA breakthrough data set then simulated by varying D_p until a best fit of the model output to the experimental data is obtained. The pore diffusivity, D_p, is incorporated here in terms of an impedance factor, τ, using the relationship $D_p = D_l / \tau$, where D_l is the free liquid diffusivity of the solute.

External Film Transfer

The parameter values resulting from model calibrations of SBA data are also presented in Table 3. A statistically significant decrease in the external film transfer coefficient with increasing preloading was observed with k_f values dropping from 3.5 x 10^{-3} to 1.0 x 10^{-3} cm/s over four weeks of preloading. These findings suggest that preloading may impact diffusion processes on the surfaces of GAC particles. One possible explanation may be that the local fluid viscosity of the film surrounding a particle is increased by the presence of the relatively large molecular weight background organic matter preloaded on the particle surface, resulting in decreased rates of diffusion of TCE across the hydro- dynamic boundary layer. Another possibility is that some effective area for TCE mass-transfer flux into the carbon particle is lost due to a partial blockage of external pores by preloaded background organic matter.

Intraparticle Diffusion

The values measured for the pore diffusion impedance factor, τ, can be related to changes in resistance to TCE diffusion with preloading. Impedances near unity imply that the total mass diffusion is on the order of that given by the free liquid diffusivity of the solute, i.e., the carbon pore spaces are relatively open and free of obstruction. Impedances several factors greater than unity indicate that intraparticle diffusion of TCE is greatly restricted by the presence of preloaded background organic matter in the carbon pores. The impedance factors presented in Table 3 can be interpreted within the above framework. For unpreloaded carbon, the τ value was 1.2. This indicates that significant intraparticle transport of TCE occurred in the unpreloaded carbon. For all of the Huron River water preloaded carbons, however, the impedance factors were found to be significantly greater than unity. Values of τ between five and ten indicate that intraparticle diffusion was 5-10 times slower than that for TCE in pure water. This suggests that preloaded background organic matter posed a significant resistance to TCE diffusion within the carbon pores. The high upper confidence limits of τ values greater than three was caused by an increasing insensitivity of the model to this parameter as intraparticle impedance became large.

Time-Variable Parameter Model Sensitivity

To assess the relative contribution of each isotherm and rate parameter to the overall shape of the breakthrough profile in the time-variable parameter modeling method, each parameter was varied individually from a constant baseline condition. The baseline case was produced by simulating breakthrough in the bench-scale fixed-bed adsorber using unpreloaded carbon isotherm and rate parameters based on the CMBR isotherms and corresponding SBA rate parameters for Huron River water over a period typical of the run times used in this research. For the time-variable parameter simulations, a nonlinear regression software package (SYSTAT, Systat, Inc., Evanston, IL) was used to obtain continuous functions of the isotherm and rate parameter values with preloading time that were incorporated into the model. Exponential and linear expressions, which give parameter values as functions of pre- loading time, were fitted to the isotherm and rate data given in Table 3. The resulting equations are presented in Table 4. In the sensitivity analysis, parameter expressions for the Huron River water CMBR-based system were varied one at a time from the baseline condition, and the resulting simulations are plotted in Figure 8.

Figure 8. Time-Variable Modeling Sensitivity to Input Parameters (Carter and Weber, 1994)

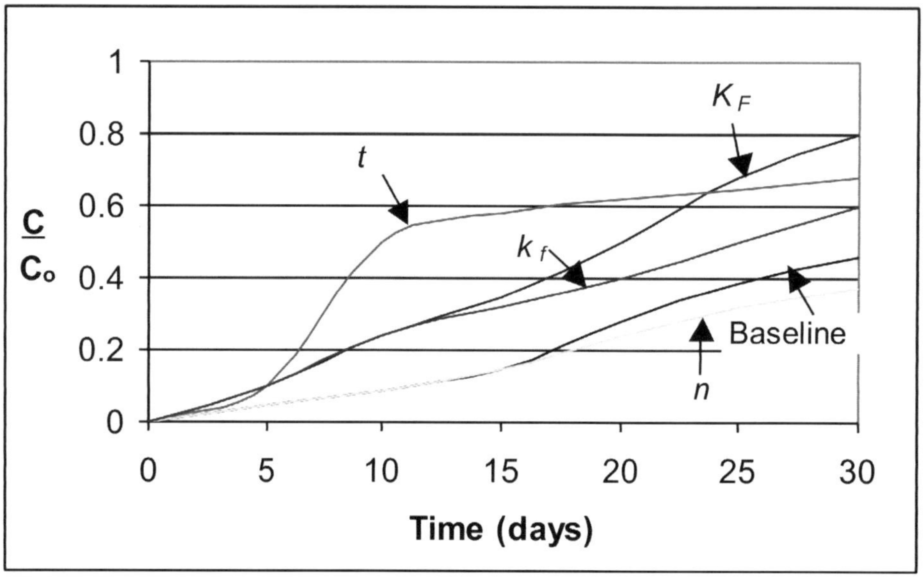

Examination of the breakthrough curves presented in Figure 8 reveals that the greatest impact on this profile was caused by changes in Freundlich K_F and impedance factor values. Decreases in the Freundlich K_F value caused a sharp rise in the latter portion of the breakthrough profile. This rapid rise was probably due to the coupled effect of an accelerated approach to equilibrium as the carbon became loaded with solute and a simultaneous lowering of capacity as K_F values declined. The effect produced by increasing impedance largely impacted intermediate time data. The decreasing rate of uptake of solute by the carbon as τ increased promoted faster breakthrough of solute. Smaller impacts were noted for variations in Freundlich n and film transfer coefficient (k_f) values. Increases in n actually produced a decrease in the breakthrough profile in the sensitivity analysis. This result is an artifact arising from changing τ while holding K_F constant which, in effect, results in an apparent increase in isotherm capacity. In actuality, increases in n are always accompanied by decreases in K_F. The notable feature of the curve resulting from the n variation is that the effect is relatively small, although it still would provide some offset to the impacts of declining K_F values. The range of impact of decreasing film-transfer coefficient (k_f) values is surprising; changes in this parameter are significant over the entire period simulated. The mechanistic nature of this parameter would suggest that k_f variations would primarily impact early time data and diminish thereafter (Weber and Liu, 1980). This behavior may be due to the relatively large drop in k_f (~70%) that was seen with these data. It may also relate to the portion of the breakthrough curve simulated, which reflected only about 50 % breakthrough. As breakthrough approaches completion, the impact of changes in k_f may very well decrease.

Fixed-Bed Adsorber Experimental Data and Modeling Results

Breakthrough data were obtained for TCE in the background water over a period of four weeks using the bench-scale fixed-bed adsorber system. For each set of fixed-bed breakthrough data, both constant parameter and time-variable parameter simulations were made using the pore diffusion model. For the constant parameter simulations, parameter sets obtained at each of the individual preloading levels were used in the pore diffusion model to simulate bench-scale fixed-bed adsorber TCE breakthrough behavior. The time-variable parameter pore diffusion model was used to ascertain whether better predictive capability could be achieved if equilibrium and rate parameters are made to vary with time of operation.

The data and the constant parameter predictions for TCE breakthrough in the Huron River system are presented in Figure 9. The data are bracketed by predictions from the unpreloaded carbon and two-week preloaded parameters, with the unpreloaded carbon set over predicting removal and the two-week set under predicting removal. The close proximity of the four-week CMBR set to the two-week set prediction in Figure 9 suggests that preloading by the Huron River water was beginning to level off. Predictions resulting from use of the time-variable parameter expressions are shown in Figure 10. The predicted curve exhibited a sharper breakthrough profile, with more removal of TCE at early times.

Figures 9. Constant Parameter Predictions and Data for TCE Breakthrough (Carter and Weber, 1994)

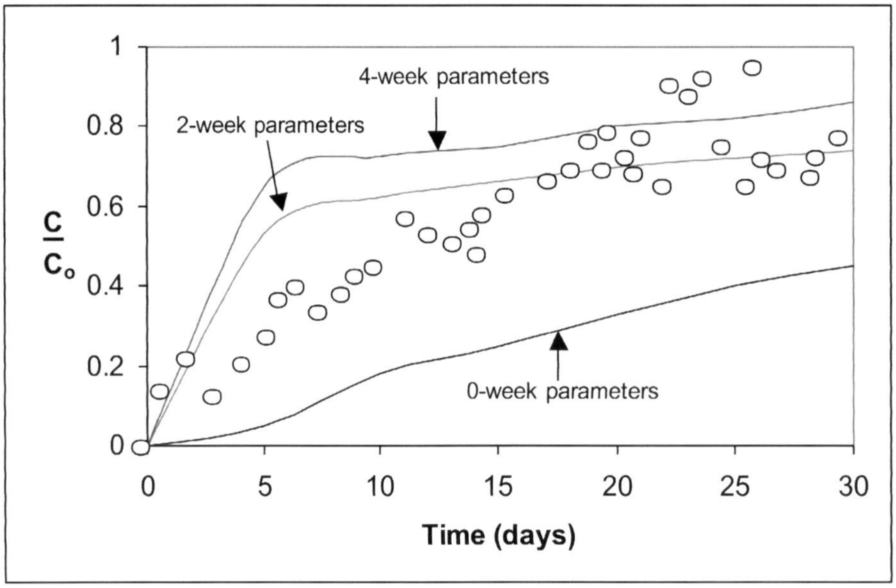

Figure 10. Time-Variable Model Predictions and Data for TCE Breakthrough (Carter and Weber, 1994)

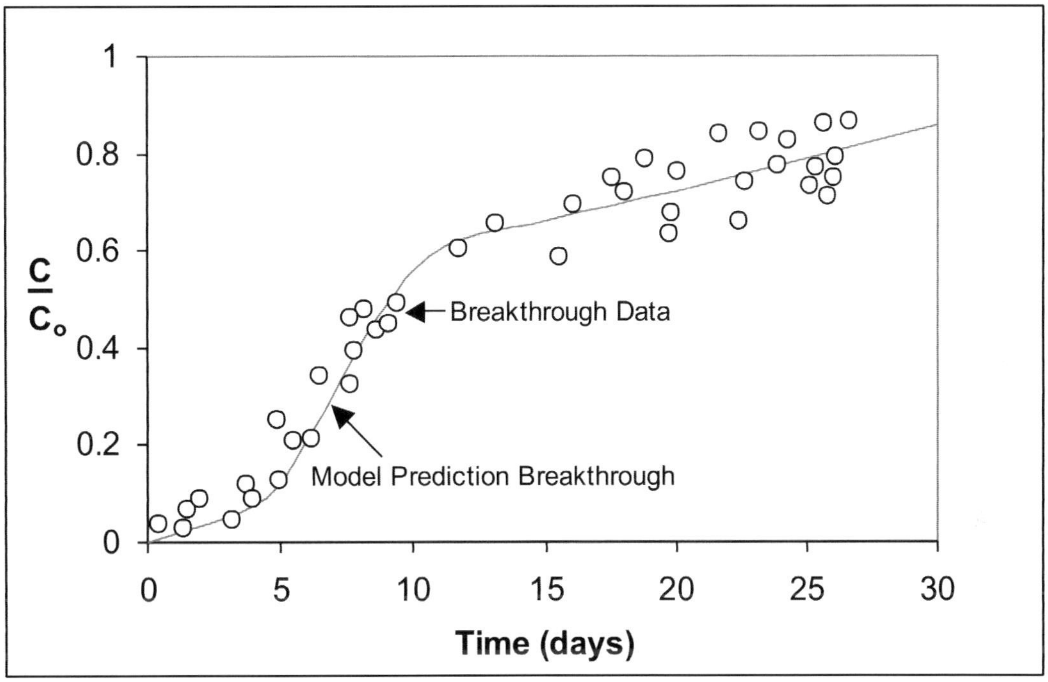

The sensitivity analysis presented earlier in Figure 8 can provide insight into the reasons why a time-variable parameter approach better fits the data. Intraparticle pore diffusion resistance was shown to primarily influence intermediate time data, causing a rapid breakthrough over this region, while decreasing capacity (as measured by K_F) significantly impacted later time data. Individually, the unpreloaded carbon parameters estimated too much capacity and intraparticle mass transport, while the two-week parameters gave levels that were too much reduced. With the time- variable parameter model, the impacts due to preloading appear to be apportioned over the run time in a manner representative of the changes in equilibrium capacity and rates of adsorption that occurred in the bed. Thus, the transition from low to high values for τ caused an acceleration of breakthrough due to higher mass-transfer resistances when capacity was high over intermediate time data. At later times, when intraparticle resistance reached a limiting value, the enhanced upward trend of the breakthrough data was continued as the K_F capacity effects became more significant.

For systems in which adsorption rate and equilibrium parameters do not change markedly over run time, fixed-bed adsorber breakthrough data may be adequately simulated using a constant-parameter modeling approach and some representative "average" parameter set.

Indeed, fixed-bed adsorber removal of target organic compounds in a background organic matter matrix has been successfully modeled using a single set of isotherm and rate parameters obtained on unpreloaded carbons (Smith and Weber 1989, Summers et al 1989, and Speth and Miltner 1989). These studies either used concentrations of target compounds at least an order of magnitude higher than those used in this work or were conducted over significantly shorter periods of time (a few days). Enhanced target compound adsorption rates from higher concentrations or a reduced time for preloading from the shorter run times would reduce the scope of any parameter changes which might occur, resulting in reasonably accurate fixed- bed adsorber modeling with a single parameter set. Conversely, in the time-variable-parameter modeling approach necessary for accurate predictions of TCE breakthrough data for the Huron River water, parameters varied a great deal over the period of study. Thus, a time-variable parameter modeling approach may be beat suited when target compound concentrations are low relative to background organic matter and fixed-bed adsorber run times are long enough for significant preloading to occur.

Conclusions and Practical Implications

The HRW case study breakthrough data summarized in the preceding section observed to lie between model simulations based on parameters derived from unpreloaded carbon data and those based on preloaded carbons. This suggests that, in practice, competitive sorption posed by background organic matter to a target solute will not be strictly simultaneous or strictly preloading in nature. For background waters that do not cause extensive changes in parameter values with preloading, the use of parameters obtained on unpreloaded or an intermediately preloaded carbon may reasonably simulate fixed-bed adsorber behavior. One indication of this state may be that intraparticle diffusion in a preloaded carbon is not yet substantially diminished from the free liquid diffusion of the target solute. In cases where isotherm and rate parameters are observed to change with pre- loading over the time period of interest, the time-variable parameter approach is the likely candidate for accurate predictions. An intraparticle diffusion coefficient several factors less than free liquid diffusion of the target solute may be a sign for this condition.

The tendency observed in the Huron River water data for preloading effects to approach limiting values suggests a third modeling regime; i.e., use of one set of constant parameters representative of a highly preloaded state. This approach was investigated by simulating breakthrough from a one-meter long column over a period of one year by both time-variable and constant parameter methods, using parameters representative of the preloading effects found in the Huron River treatment system. In the time-variable simulation, parameters were allowed to vary over the first four weeks of preloading in the same fashion as for the shorter time predictions discussed earlier. After four weeks of simulation, the values were fixed at those levels. In the constant parameter approach, the parameters were set at the four-week preloading level over the entire simulation time. A breakpoint of 10% of influent concentration was chosen for comparison purposes. The constant parameter modeling approach gave a time to breakpoint of 107 days while the time-variable parameter simulation predicted 143 days. The "extra" capacity afforded by the time-variable parameter simulation over the first four weeks of preloading (i.e., before the model parameters became equivalent) is surprisingly large given the one-year simulated run time. Nonetheless, use of a constant parameter approach with parameters representative of an extensively preloaded state would, in practice, provide a conservative estimate of overall adsorber capacity that would be acceptable for preliminary design calculations.

While three possible modeling regimes have been proposed, the uniqueness of each adsorption system precludes determination of which predictive approach is most appropriate without performing some isotherm and rate parameter measurements. Because of the heterogeneous characteristics of background organic matter, the amount of TOC preloaded onto a given mass of carbon is not a reliable indicator of the resulting preloading effect on adsorption of a target compound. Moreover, characteristics of the solute itself will influence the impact of preloading on its adsorption equilibria and rates. The impact of preloading may be more severe for a weakly hydrophobic compound such as TCE than for a highly hydrophobic solute that would tend to sorb to preloaded organic matter. Exact identification of the mechanisms of the preloading process, assessment of key properties that can be used to quantify the preloading potential of a background organic material, and an understanding of how solutes interact with preloaded organic matter in the preloaded state are areas requiring further research.

Acknowledgements

This paper draws on an invited presentation by the author at an International Water Association (IWA) sponsored Workshop on Biological Activated Carbon Filtration held at Delft University of Technology in the Netherlands nine years ago. Much more significantly, the material described in the paper is drawn from an extensive series of investigations conducted by former students of the author, as evident in the many literature citations thereto. All credits for any such intrinsic value this "capstone" paper may offer is due them. Accountability for flaws and/or shortcomings associated with this effort to compile and describe the overall results of their work accrues on the other hand solely to the author.

References

Carter, M.C., Weber, W.J. Jr., and Olmstead, K.P. (1992) "Effects of Background Dissolved Organic Matter on TCE Adsorption by GAC" *Journal of the American Water Works Association, 84*, 8, 81-91

Carter, M.C., Weber, W.J. Jr., (1994) "Modeling Adsorption of TCE by Activated Carbon Preloaded by Background Organic Matter" *Environmental Science and Technology*, 28, 4, 614-623

Crittenden, J.C., Hutzler, N.J., Geyer, D.G., Oravitz, J.L., Friedman, G. (1986) "Transport of Organic Compounds With Saturated Groundwater Flow: Model Development and Parameter Sensitivity" *Water Resources Research*, 22, 271-284

Karanfil, T, Kilduff, J.E., Schlautman, M.A., and Weber, W.J.Jr (1996) "Adsorption of Organic Macromolecules by Granular Activated Carbon. 1. Influence of Molecular Properties under Anoxic Solution Conditions" *Environmental Science and Technology* 30, 7, 2187-2194

Karanfil, T, Kilduff, J.E., Schlautman, M.A., and Weber, W.J.Jr (1998) "The Oxygen Sensitivity of Organic Macromolecule Sorption by Activated Carbon: Effects of Solution Chemistry" *Water Research*, 32, 1, 154-164

Smith, E.H. and Weber, W.J. Jr. (1989) "Evaluation of Mass Transfer Parameters for GAC Adsorption from Complex Organic Matrices" *Environmental Science and Technology*, 23, 713-722

Speth, T.F. and Mitner, R.J. (1989) "Effect of Preloading on the Scale-up of GAC Microcolumns" *Journal of the American Water Works Association*, 81, 4, 141-148

Summers, R.S., Haist B., Koehler, J., Ritz, J., Zimmer, G. and Santheimer, H. (1989) "The Influence of Background Organic Matter on GAC Adsorption" *Journal of the American Water Works Association*, 81, 5, 66-74

Weber, W.J.Jr. and DiGiano, F.A. (1996) *Process Dynamics in Environmental Systems*, Wiley-Interscience, New York

Weber, W.J.Jr., and Liu, K.T. (1980) "Determination of Mass Transport Parameters for Fixed-Bed Adsorbers" *Chemical Engineering Communications*, 6, 49-60Weber, W.J. Jr., Katz, L.A., and McGinley, P.A. (1992) "A Distributed Reactivity Model for Sorption by

Soils and Sediments: 1. Conceptual Basis and Equilibrium Assessments" *Environmental Science and Technology*, 26, 10, 1955-1962 (1992)

Nomenclature

n	=Freundlich isotherm exponent	[-]
K_F	=Freundlich isotherm capacity factor	[*]
$K_{F,M}$	=Modified Freundlich isotherm capacity factor	[**]
k_f	=film mass transfer coefficient	[cm/s]
C_e	=equilibrium solution-phase concentration	[mg or µg/L]
D_o	=initial sorbent dosage	[g/L]
D_p	=pore diffusion coefficient	[cm^2/s]
D_l	=free liquid diffusion coefficient	[cm^2/s]
q_e	=equilibrium solid-phase concentration	[mg or µg/g]
$\phi(x)$	=function of x	
τ	=diffusion impedance	

*Units dependent upon units of q_e and C_e
**Units dependent upon units of q_e, C_e and D_o

Abbreviations

CMBR	=completely mixed batch reactor
FBR	=fixed bed reactor
GAC	=granular activated carbon
HRW	=Huron River water
NOM	=natural organic matter
ppb	=parts per billion
ppm	=parts per million
SBA	=short bed adsorber
TCE	= trichloroethylene
TOC	=total organic carbon

(115b) Water Challenges and Opportunities In the Emerging World

Adil Dhalla, GE Singapore Water Technology Centre at NUS, Singapore, Singapore

The human population has grown more in the last 50 years than it did in the previous 4 million years. We can see this as both a challenge and an opportunity. Two of our key resources – power and water, sustain the quality of our life, and the renewing of these is economically sound. Less than 1% of the world's freshwater is readily available for human use. The amount of water that's safe for us to use is declining sharply due to pollution and contamination. We need water not only to survive, but also to prosper and grow. With water demand tripling by 2030, we have to be three times as smart. Our business, GE Water & Process Technologies brings together experienced professionals and advanced technologies to solve the world's most complex challenges related to water availability and quality, increased productivity and cost reduction, and environmental regulations.

See more of this Session: Plenary Session on Water Technology Challenges and Advances (invited speakers)
See more of this Group/Topical: Topical 1: Water Technology for Developed and Developing Countries (see also Separations Division)

(115c) Epa's Research and Development Program for Innovative Water Technologies

Thomas Speth, National Risk Management Research Laboratory, USEPA, Office of Research and Development, Cincinnati, OH

EPA's Research and Development Program for Innovative Water Technologies

Thomas F. Speth

The EPA's Administrator Lisa Jackson has released a drinking water strategy to protect public health from contaminants in drinking water. An element of this strategy is to foster development of new drinking water technologies to address health risks posed by a broad array of contaminants. Specifically, EPA wishes to 1) develop, identify, and evaluate drinking water systems of the future that have minimal carbon and water footprints and are more sustainable than current practices; 2) conduct strategic demonstration projects with small communities to address a broad suite of contaminants while maintaining safe drinking water through their distribution systems at reasonable and predictable costs; 3) develop infrastructure management that accommodates the new generation of technologies and monitoring strategies. As part of this effort, the EPA along with the Small Business Administration are catalyzing the formation of a water technology innovation cluster that involves industry, government, academia, and non-government organization partners working together to bring novel technologies to the drinking water market. Emphasis is being placed on systems-based approaches that will utilize and benefit from integrated trans-disciplinary research. This presentation will address the rationale for the program and the efforts that are currently underway.

See more of this Session: Plenary Session on Water Technology Challenges and Advances (invited speakers)
See more of this Group/Topical: Topical 1: Water Technology for Developed and Developing Countries (see also Separations Division)

(115d) Water Treatment Opportunities Utilizing the Dramatic Nano-Fluidic Properties of Carbon Nanotube Membranes

Bruce J. Hinds, Chemical and Materials Eng., University of Kentucky, Lexington KY, KY

Carbon nanotubes have three key attributes that make them of great interest for novel membrane applications 1) atomically flat graphite surface allows for ideal fluid slip boundary conditions 2) the plasma cutting process to open CNTs inherently places functional chemistry at CNT core entrance and 3) CNT are electrically conductive allowing for electrochemical reactions and application of electric fields gradients at CNT tips. In general, the transport mechanisms through CNT membrane are a) ionic diffusion is near bulk expectation with no enhancement from CNT b) gas flow is enhanced by ~1-2 order of magnitude due to specular reflection off of flat graphitic surface c) and pressure driven flux of a variety of solvents are 4-5 ORDERS OF MAGNITUDE FASTER than conventional Newtonian flow due to atomically flat graphite planes inducing nearly ideal slip conditions. Facilitated transport, electrocatalytic defouling and electroosmotic pumping through CNTs are observed. For water treatment applications such, as RO desalination, a 2 wt% CNT polymer has the potential to result in a 100 fold reduction in membrane area. However one of the key difficulties is that to impart membrane chemical selectivity, chemical functionalization of CNT surfaces ruin the slip boundary condition and dramatically reduce flow enhancements. Efforts in the field generally rely on placing high charge density at CNT tips to reject ions or to precisely control CNT diameter to allow only single file columns of water.

See more of this Session: Plenary Session on Water Technology Challenges and Advances (invited speakers)
See more of this Group/Topical: Topical 1: Water Technology for Developed and Developing Countries (see also Separations Division)

(185a) A Perspective On the Impact of Material Sciences Innovations and Sustainable Development Considerations On Future Membrane Developments In the Water Industry

Herv Buisson, Solutions and Technologies Americas, Veolia Water, Cary, NC

The early membrane developments from key pioneers and innovators like Pierre have had a profound impact on the water industry over the past 50 years or so. My presentation will first recognize Pierre's outstanding contribution to the field, seen from the small window of my personal interactions with him over the years.

Our world is fast changing in many ways, and needs and opportunities for further membrane innovations in our industry are plenty. I will use practical examples of recent projects and developments done in our company to illustrate how the future of membrane technology is being shaped by two drivers: a technology-push driver fueled in part by recent innovations in material sciences, and a market-pull driver where water footprint and sustainable development considerations are playing an increasingly important role.

As chosen case studies to illustrate those points, I have selected to spend some time in my presentation on our Anaerobic MBR recent developments , as well as on water-footprint considerations as they may apply to membrane processes.

See more of this Session: Session In Honor of Dr. Pierre Cote: Membrane Separations (Invited Papers)
See more of this Group/Topical: Topical 1: Water Technology for Developed and Developing Countries (see also Separations Division)

(185b) ZW500 MBR: 15+ Years of Evolution

Geert-Henk Koops, Water & Process Technologies, GE Power & Water, Oakville, ON, Canada

Over 15 years ago GE Water & Process Technologies (in those days Zenon Envorinmental) launched a revolutionary submerged hollow fiber product for the separation of biologically cleaned water and the mixed liquor in an application called MBR. Over the years this product has seen a number of modifications and a spectacular reduction of energy consumption, specifically in membrane scouring.

During this presentation I would like to take you on a journey of product development, highlight 3 game changers in this product line and discuss how new developments are validated within our business.

See more of this Session: Session In Honor of Dr. Pierre Cote: Membrane Separations (Invited Papers)
See more of this Group/Topical: Topical 1: Water Technology for Developed and Developing Countries (see also Separations Division)

(185c) Gas Sparging for Membrane Fouling Control

Pierre P. Brub, Civil Engineering, University of British Columbia, Vancouver, BC, Canada

Membrane systems are increasingly being used in water and wastewater treatment applications because of their ability to effectively retain particulate and soluble contamiants of concern. However, some of the retained material (i.e. foulants) can accumulate on or in the membrane. Unless controlled, this accumulation can significantly increase the resistance to the permeate flow through the membrane, and result in an increase in the capital and operating costs for the system.

Gas sprging is commonly used in submerged membrane systems to limit the accumulation of foulants at the membrane surface. The effects of gas sparging on the mass transfer of foulants at membrane surfaces are reviewed, and the sparging conditions that minimize the accumulation of foulants are presented. Emphasis is placed on the effect of the membrane system configuration and sparging scenario as they can significantly affect the power costs associated fouling control. This is of particular importance since the power costs associated with gas sparging can account for up to 40% of the operating cost of a membrane system.

See more of this Session: Session In Honor of Dr. Pierre Cote: Membrane Separations (Invited Papers)
See more of this Group/Topical: Topical 1: Water Technology for Developed and Developing Countries (see also Separations Division)

(185d) Fouling Causes and Control Strategies for Membrane Bioreactors In Municipal Wastewater Treatment

Hongde Zhou, School of Engineering, University of Guelph, Guelph, ON, Canada

This paper summarizes the extensive experimental results over last six years using large-scale membrane bioreactor pilot plants to treat municipal wastewater. First, the causes of membrane fouling were identified by monitoring the fouling rate under different operating conditions and characterizing the fouling materials deposited on the membrane surface. Different mechanisms were demonstrated between short-term physicochemical and long-term irreversible membrane fouling. Colloidal particles represented by cTOC were identified as a fraction of mixed liquor controlling the critical flux and short-term membrane fouling, whereas the irreversible fouling was initiated by depositing the proteins directly on membrane wall, followed by the accumulation of carbohydrates to hold bacteria tightly. Secondly, new fouling control strategies will be discussed, with emphasis on their effectiveness and energy consumption. Finally, selected case studies will be used to illustrate the applications of these fouling strategies in practice.

See more of this Session: Session In Honor of Dr. Pierre Cote: Membrane Separations (Invited Papers)
See more of this Group/Topical: Topical 1: Water Technology for Developed and Developing Countries (see also Separations Division)

(185e) Ozonation of a Non-Nitrified Secondary Effluent to Reduce Microfiltration Fouling

R. Shane Trussell, Trussell Technologies, Inc, San Diego, CA

Microfiltration (MF) is a common unit process for treating municipal wastewater effluent for indirect potable reuse. However, effluent organic matter (EfOM) can foul MF membranes thereby reducing flux, increasing cleaning frequencies, and shortening membrane longevity. Research has shown high molecular weight (MW) organics are the primary source of MF fouling, and ozonation is known to preferentially oxidize large MW organics. However, this application of ozone is uncommon and there are few references discussing it in the scientific literature. Therefore, pilot testing was performed to evaluate the feasibility of oxidizing EfOM with ozone to restore MF performance while treating non-nitrified secondary effluent with poor water quality. Two parallel MF treatment trains, one with ozone pretreatment (O_3-MF) and the other without it, were used to compare the benefits of EfOM oxidation with ozone. The MF unit without ozone pretreatment operated at 19.3 gfd for < 3 days before it shutdown from high transmembrane pressure, but the MF unit with ozone pretreatment (6 mg/L transferred dose) operated at 20 gfd for 21 days. These results demonstrated ozonation of EfOM extended the interval between MF cleanings by seven-fold, indicating this MF pretreatment strategy could provide significant benefits if implemented at full-scale.

See more of this Session: Session In Honor of Dr. Pierre Cote: Membrane Separations (Invited Papers)
See more of this Group/Topical: Topical 1: Water Technology for Developed and Developing Countries (see also Separations Division)

(185f) Future Directions for Wastewater Treatment

Pierre Cot, COTE Membrane Separation Ltd., Hamilton, ON, Canada

This presentation will review future directions for wastewater treatment.

See more of this Session: Session In Honor of Dr. Pierre Cote: Membrane Separations (Invited Papers)
See more of this Group/Topical: Topical 1: Water Technology for Developed and Developing Countries (see also Separations Division)

(270a) Nanomaterials and Recent Developments In Membrane Technology In the Middle East

Suzana Pereira Nunes, Water Desalination and Reuse Center, KAUST, Thuwal, Saudi Arabia

The presentation will summarize recent developments of membrane technology in the Middle East, including applications and emerging research activities. Initiatives related to the new Water Desalination and Reuse Center at KAUST, Saudia Arabia, will be emphasized, discussing new materials and morphology control in membranes.

See more of this Session: Water Technology Advancements In Asian Countries
See more of this Group/Topical: Topical 1: Water Technology for Developed and Developing Countries (see also Separations Division)

(270b) Trend of Membrane Technology for Water Treatment In Japan

Hideto Matsuyama, Kobe university, Center for Membrane and Film Technology, Kobe, Japan

Water shortage is one of the serious environmental problems which must be solved immediately. Membrane technology is widely applied for water treatment and receiving more attention since it is an outstanding process. In Japan, membrane technology has been attracting much attention. Share of Japanese companies in total membrane production in the world is reported to be about 60%. Especially in the RO membrane market, the share is as high as 70%. Thus, Japanese companies are very active in the membrane production field.

In this situation, Japanese government is eager to support the membrane researches for water treatment. Several projects are now going on. For example, in Core Research for Evolutional Science and Technology (CREST) project, the research area of "Innovative Technology and System for Sustainable Water Use" was selected. The target of this research area is development of innovative technologies for realizing sustainable water management by mitigating water problems intensified by climate change. Many researches about membrane technology for water treatment have been adopted in this program. Another project is "Mega-ton Water System" in Funding Program for World-Leading Innovative R&D on Science and Technology (FIRST Program). The target of this project is development of RO membrane system in the scale of mega-ton/day. The projects about innovative membrane have been started also in New Energy and Industrial Technology Development Organization (NEDO).

In academia, the first membrane center was established in Kobe University in 2007. Now faculty member is 15 and total students number is over 100. The member of companies is mote than 40. One of the main research targets in this membrane center is water treatment by the membrane. Many co-operation projects between the center and the companies have been progressing.

See more of this Session: Water Technology Advancements In Asian Countries
See more of this Group/Topical: Topical 1: Water Technology for Developed and Developing Countries (see also Separations Division)

(270c) Water Treatment R&D In Korea

Seockheon Lee, Water Research Center, Korea Institute of Science and Technology, Seoul, South Korea

Water scarcity is common problem in countries. Industrial development, population growth and urbanization have been accelerated water consumption. Global scale climate change makes the water shortage problem more severe. Water reuse can be a reliable alternative to supplying water for domestic and industrial purpose. Membrane filtration implemented technology is increasingly expanding its application to water and wastewater treatment in Korea. New water treatment plants are introducing MF and UF as core processes. Reverse osmosis is replacing distillation for seawater desalination, and considered as the final barrier to remove harmful contaminants from waste water to reuse. R&BD on water treatment in Korea will be introduced. Current research on water treatment of KIST will be also covered.

See more of this Session: Water Technology Advancements In Asian Countries
See more of this Group/Topical: Topical 1: Water Technology for Developed and Developing Countries (see also Separations Division)

(270d) Applications of Membrane Technologies for Water Treatment In Taiwan

Da-Ming Wang, Chemical Engineering, National Taiwan University, Taipei, Taiwan, Kuo-Lun Tung, R&D Center for Membrane Tech. and Dept. of Chem. Eng, Chung Yuan Christian University, Chung Li, Taiwan and Juin-Yih Lai, R & D Center for Membrane Technology and Department of Chemical Engineering, Chung Yuan University, Chung-Li, Taiwan

Though Taiwan's annual rainfall is about 3.5 times more than the average annual rainfall of the world, Taiwan is still categorized by the United Nations with water shortage problems. The main reason is the uneven rainfall in Taiwan: nearly 80 percent of precipitation falls in between May and October, and water is difficult to retain in rivers because of the dramatic topographical changes. To sustain the water supply in Taiwan, several membrane desalination and water-treatment units have been set up in recent years. Also, increasing research interest has been drawn to this topic. In this presentation, we will discuss the recent applications of membrane technologies in desalination, drinking-water treatment, and wastewater treatment in Taiwan.

See more of this Session: Water Technology Advancements In Asian Countries
See more of this Group/Topical: Topical 1: Water Technology for Developed and Developing Countries (see also Separations Division)

Green Preparation of polyvinylidene fluoride Microfiltration Membranes Via Thermally Indueced Phase Separation Method Using Diphenyl Carbonate and Diphenyl Ketone As Diluents

X. L. Wang, J. Yang, Y. K. Lin, Y. Tian, Y. H. Tang
Beijing Key Laboratory of Membrane Materials and Engineering, Department of Chemical Engineering,
Tsinghua University,Beijing 100084, P.R. China
xl-wang@tsinghua.edu.cn

Extended Abstract

Diphenyl carbonate (DPC) and diphenyl ketone (DPK) were used as diluents to prepare microporous polyvinylidene fluoride (PVDF) membranes via thermally induced phase-separation (TIPS) method. The liquid - liquid phase-separation phenomena were found in PVDF/DPC and PVDF/DPK systems during the thermally induced phase-separation processes. The monotectic points of PVDF/DPC and PVDF/DPK systems appeared at PVDF concentration approximately 30 and 56 %(wt), respectively. This suggested that PVDF membrane with a bicontinuous or cellular structure could be obtained when the PVDF concentration was less than the point values. The effects of polymer concentration and quenching temperature on the membrane structure, porosity, and tensile strength were also investigated. The results showed that low polymer concentration and high quenching temperature led to a large pore size membrane. The tensile strength of the membrane increased with the increase of the polymer concentration, but the porosity of the membrane decreased. On condition that the PVDF concentration was higher than monotectic points, only polymer crystallization occurred and a typical compact spherulitic structures were obtained.

Moreover, a binary diluent of DPK and secondary solvent such as 1,2-propylene glycol (PG) was used to replace the single diluent in TIPS process. The phase diagram for the PVDF/binary diluent of DPK and PG system was measured in the range of the PG/DPK mass ratio changing from 0 to 2/3. Then the effects of the PG/DPK mass ratio and the PVDF concentration on membrane cross-section structures and tensile strength were also investigated. The results showed that the addition of PG brought about a shift of the cloud point curve to a higher temperature and the extension of the liquid-liquid phase separation region to a higher polymer concentration. Therefore a bicontinuous cross-section structure was obtained when the PG/DPK mass ratio was 3/7 and the polymer concentration was 30 wt%. As an increase of the PG/DPK mass ratio, the tensile strength increased gradually at a fixed PVDF concentration. Moreover, for the same PG/DPK mass ratio, the cross-section microstructure changed from a bicontinuous or a cellular structure to a spherulitic structure, and the tensile strength increased drastically as the polymer concentration increased from 20 wt% to 50 wt%.

According to the properties of diluents, the industrial scale production line with a twin-screw extruder of microporous PVDF hollow fiber membranes was established and a series of different type hollow fiber membranes were prepared for the corresponding applications of pressure-driven modulus, submersed modulus and membrane bio-reactor modulus. At the same time, the full recycle of diluents was almostly actualized by using a series of unit opertaions such as extraction, distillation, delamination, sedimentation and drying processes. It means there is no waste emission in the preparation of PVDF membranes or it is a green process.

Acknowledgment

This work was supported by National Basic Research Program of China (2009CB623404) and Key-Project of Beijing Municipal Science & Technology Commission (D0406003040191).

References

1. Jian Yang, Dawei Li, Yakai Lin, Xiaolin Wang, Formation of a Bicontinuous Structure Membrane of Polyvinylidene Fluoride in Diphenyl Ketone Diluent via Thermally Induced Phase Separation [J], Journal of Applied Polymer Science, 2008, 110(1): 341-347
2. Yakai Lin, Yuanhui Tang, Xiaolin Wang. Formation of a Bicontinuous Structure Membrane of Polyvinylidene Fluoride in Diphenyl Carbonate Diluent Via Thermally Induced Phase Separation[J]. Journal of Applied Polymer Science, 2009, 114, 1523-1528
3. Yuanhui Tang, Yakai Lin, Wenzhong Ma, Ye Tian, Jian Yang, Xiaolin Wang, Preparation of Microporous PVDF Membrane via Tips Method Using Binary Diluent of DPK and PG[J]. Journal of Applied Polymer Science, 2010, 118(6): 3518~3523

(270f) Water Sustainability and R & D In Singapore

Tai-Shung Chung, Chemical & Biomolecular Engineering, National University of Singapore, Singapore, Singapore

Water, energy, affordable healthcare and global warming are four major concerns globally resulting from clean water shortages, resource depletion, high costs of pharmaceuticals, and changing climate conditions. Water reuse and seawater desalination are especially important for Singapore because we are an island state with very limited water supply. Up to 50% of water consumed in Singapore is purchased from a neighboring country. Over the past decade, water demand in Singapore increased at approximately 3% per annum. This is expected to rise considerably because of the rapid expansion of our city and local industries. A review on water sustainability and R & D at Singapore will be carried out. We will also cover membrane research at NUS for water sustainability.

Email: chencts@nus.edu.sg; Website:1) http://www.chee.nus.edu.sg/people/faculty_chungneal.html; 2) http://www.chbe.nus.edu.sg/membrane

See more of this Session: Water Technology Advancements In Asian Countries
See more of this Group/Topical: Topical 1: Water Technology for Developed and Developing Countries (see also Separations Division)

(301a) Opportunities for Government to Influence Technology Innovation In the Water Industry

Jeffrey Q. Adams, National Risk Management Research Laboratory, US EPA, Cincinnati, OH

Opportunities for Government to Influence Technology Innovation in the Water Industry

Jeffrey Adams, USEPA NRMRL, WSWRD, WQMB

This presentation will discuss recent activities initiated by EPA's National Risk Management Research Laboratory to collaborate with numerous organizations including Federal and State regulatory and permitting agencies, water utilities & communities, water treatment manufacturers & vendors, engineering/science community, and business/venture capital sectors to promote technology innovation and markets focusing on the drinking water industry.

Ongoing environmental science research provides guidance for the development Federal and State water supply policies and regulations which become major industry drivers for what role certain environmental services and technology applications will play in the water supply business and markets. The Federal government often sets the initial framework for defining water policy, issues, and needs. This presentation will explore some scenarios on what government can do with limited resources to influence and promote technology innovation and help streamline the complex technological and institutional hurdles that must be confronted as new emerging tools and water treatment applications move across the research continuum from bench concept to market-ready penetration and sustainable deployment.

On April 27, 2011, the Obama Administration released a national "Clean Water Framework" which affirms a comprehensive commitment to protecting the nation's waters. The framework emphasizes the use of partnerships among States, tribes, local governments, and diverse stakeholders to promote innovative approaches for achieving sustainable water supplies. It directs the use of the latest science and research to improve water regulations/policies and identify and address emerging pollution challenges. This complements EPA's new set of strategies for managing drinking water policy issues including: addressing water contaminants as groups as much as feasible rather than one at a time; fostering development on new drinking water technologies by utilizing private innovators and small business; using the authority of multiple statutes to protect water supplies; partner with States to develop shared access to all public water systems monitoring data.

On the forefront of consideration for regulatory groups of contaminants to manage are numerous selected carcinogenic Volatile Organic Chemicals (VOCs), several Nitrosamines, and numerous chlorination Disinfection Byproducts (DBPs). Other groups that may emerge include: Perfluorinated compounds, Organophosphate Pesticides, Carbamate Pesticides, Triazine Pesticides, Chloroacetanilides, and Cyanotoxins. Various individual chemicals receiving renewed interest include Fluoride, Chromium-VI, and Perchlorate, and various microbials and chemicals from the Contaminant Candidate List-3 (CCL3). In addition, initiatives to shift from defining drinking water MCLs from micrograms/L to nanograms/L will have an impact on regulatory policy consideration as the industry grapples with new analytical methods, how to determine health impacts at such low levels of exposure, and determining cost-effective treatment applications.

In conjunction with regulatory drivers, business market dynamics play a role in stimulating the development of innovative technology applications. In 2010, it was estimated that global venture capital investment in the water sector represented only 2% of all cleantech industry investments with the U.S. share of global investments under 30%. The estimated total U.S. water equipment market (excluding engineering/operator services and chemical consumables) was about $28 billion across drinking water, wastewater, and industrial water. The drinking water equipment market segment alone was estimated at about $12 billion, and an analysis indicated the additional available market innovation opportunity in this sector was only $1.2 billion. Factors contributing to the weak investment in water technology include the low profit potential for an industry that does not utilize full cost pricing policies for water production and distribution and the difficulties encountered with federal, state, local government regulatory, permitting, and procurement policies. The analysis indicated that facing competition from well-financed global competitors, U.S. innovators would benefit from technology test-beds and incubators, particularly those sponsored by government, providing demonstration of technology performance within the framework of current regulatory policies or that stimulate reformation of policies.

EPA's Office of Water develops policies and regulatory guidance that influences State decision-making for technology review/permitting/acceptance. Federal guidance often lists technology applications and associated treatment credits for various applications. However, if an innovative technology is developed that does not fit within the current guidance framework, it may be difficult to get States approval to use it. Though this may be an unintended consequence, State regulatory agencies are reluctant to permit technologies that do not have a credible record of performance, or technologies that are designed/configured/operated in a manner that falls outside of the expertise of a State's review staff or cannot be assessed under current federal guidance procedures. These technologies may very well be innovative and prove to be sustainable solutions down-the-road but are viewed as black-box devices that are difficult to evaluate under a current framework.

Several efforts initiated by EPA's National Risk Management Research Laboratory to promote and support technology innovation in the water industry will be discussed. These include providing technology development test-beds at EPA facilities thru equipment vendor partnerships thru CRADA mechanisms, field and lab performance testing of various types of market-ready water treatment technologies under the Environmental Technology Verification (ETV) Program, fifty full-scale demonstration studies of new and innovative arsenic removal technologies deployed at operating water utilities, early-phase testing of new technologies under the SBIR Program, collaborations with the Water

Research Foundation, and Water Environment Research Foundation, and EPA's collaboration with a Water Technology Innovation Cluster (WTIC).

These programs allow the government to act as a catalyst to support independent credible evaluation / verification of various technologies thru third-party partnerships and collaborators. Funding for testing studies varies across different programs ranging from complete government funding, partial vendor subsidy with other collaborator funding, to complete vendor cost coverage of the testing. Testing protocols have been developed thru a stakeholder process that allows feedback from regulatory, science/ engineering, water utility/user community, and technology manufacturer/vendor viewpoints. The independent performance testing results can offer many contributions to technology innovation. Results may inform future regulatory policy reform. Testing results provide manufacturers with identification of strengths and weaknesses of their specific products to foster improvements in their design and operation based on lessons learned. These programs have provided State and local government permitting and procurement staff with credible data in standardized reports to inform decision-making. Efforts have provided protocols for the basis of standardized testing in the scientific/engineering community and stimulation of NSF-type Certification programs to help streamline State approval confidence of innovative technologies. Test-bed studies on market-ready equipment have in some cases provided positive market outcomes for vendors in achieving market penetration and States permitting advantages. Under these EPA programs, study results and lessons learned have been communicated to federal and State regulatory Agency staff and various stakeholder groups thru numerous EPA sponsored Workshops and Webinars in recent years.

Under the WTIC effort, EPA's role is a catalyst to collaborate with other organizations within a regional technology innovation cluster framework in the Ohio, Kentucky, Indiana tri-state region. The Cluster is a geographic concentration of interconnected firms - businesses, suppliers, and service providers - and supporting institutions - local government, business chambers, universities, investors, and others - that work together in an organized manner to promote economic growth and technological innovation, in this case associated with the water technology sector. The mission of WTIC is to develop and commercialize innovative technologies to solve environmental challenges and spur sustainable economic development and job creation.

See more of this Session: Industry and Government Activities In Water Purification
See more of this Group/Topical: Topical 1: Water Technology for Developed and Developing Countries (see also Separations Division)

Navy Shipboard Desalination

Ian Peek, Code 985, Naval Sea Systems Command - Philadelphia, Philadelphia, PA
David Nordham, Code 923, Naval Sea Systems Command - Philadelphia, Philadelphia, PA
Paul Armistead, Office of Naval Research, Arlington, VA

Shipboard desalination is required to produce potable water on U.S. Navy ships, for personnel as well as machinery uses. The purpose of this talk is to provide an overview of Navy shipboard desalination, from distillation through legacy reverse osmosis systems to the development and deployment of future systems. A discussion of the driving forces behind the design process are included, as well as the targeted performance metrics.

Shipboard Desalination using Distillation

Prior to the introduction of reverse osmosis, shipboard desalination was accomplished using distillation. Distillation is very energy intensive, requiring approximately 424 kWh/1000gal. A significant amount of maintenance is required, shown by an operational availability of approximately 60%. Distillation units also require a large amount of chemical additions, in the form of antiscalant.

Existing Shipboard Reverse Osmosis Systems

The first RO unit on a U.S. Navy ship was installed on the *USS Fletcher* (DD 992) in 1988. Successful evaluation showed an operational availability greater than 99%. Since then, the ships have a standardized design, the Navy Standard Reverse Osmosis (NSRO), which has a 12,000 gal/day capacity. The process schematic may be seen in Figure (1).

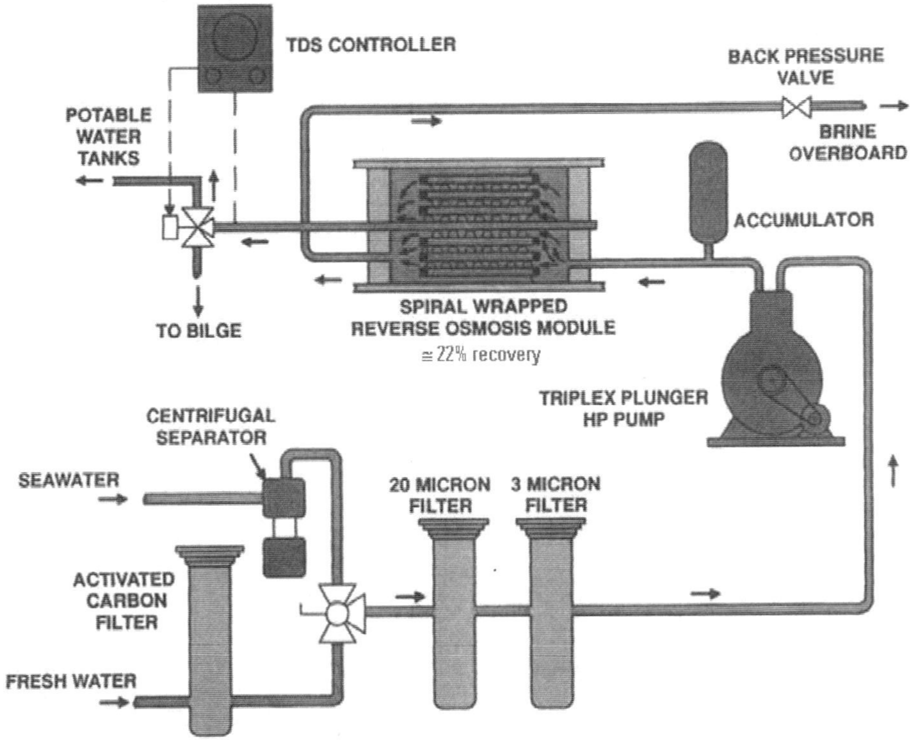

Figure (1) – Navy Standard Reverse Osmosis Process Flow Schematic

Depending on an individual ship's water requirement, the design is scaled up or down. Prefiltration is accomplished with a centrifugal strainer followed by cartridge filtration, at the 20 and 3 micron sizes. The RO section consists of a high pressure plunger pump into commercial spiral-wound membranes, after which the permeate goes to the shipboard potable water storage and the brine is discharged overboard. The system operates at 22% recovery (ratio of permeate to feed flows). Ships requiring higher purity water utilize multiple pass RO to achieve the desired level of purification.

While this is a robust and proven design, having been installed on Navy ships since the 1980s, there are several areas for improvement. The focus of the design was for deep blue sea, i.e., the open ocean. Given recent conflicts, littoral operations are more frequent. In littoral operations, the cartridge filters clog frequently, requiring dedicated manpower to replace the filter elements. This leads to additional storage requirements for new filters as well as used cartridges prior to disposal. The lifetime of cartridge filters decreased from four to six weeks in the open oceans to as little as seven hours in turbid littoral waters. The added strain on the cartridge filters increases the fouling on the RO elements, often requiring their replacement as well. Chlorination of feed seawater piping for biofouling and MIC protection raise operational concerns for RO membranes and require additional process steps for dechlorination. Lastly, the desire to increase potable water generation for improved "quality of life" for sailors cannot be met in the existing volumetric and gravimetric footprints with the NSRO design.

Advancing Technology – the EUWP Program

In 2003, the Office of Naval Research initiated a program to advance the technology used in shipboard desalination systems, the EUWP program (Expeditionary Unit Water Purification). The objectives of this program were to reduce energy requirements, minimize system size and weight, increase ship operational capabilities in the littorals, improve reliability and reduce maintenance, lower logistics requirements and lower operating and total ownership costs. The products of the EUWP program were two large scale units, with capacities of 100,000 gal/day and 300,000 gal/day, respectively.

In addition to the construction of the units, the Navy learned several critical pieces of information as a result of endurance testing. Innovative membrane prefiltration could successfully filter highly turbid coastal seawater with little maintenance requirements in 2400 hours. The use of energy recovery devices allowed for reduced energy, size, noise, and maintenance requirements of high pressure pumps. High flux RO membrane elements and hybrid configurations allowed for reduced space, weight, and energy requirements. An advanced electrodeionization device allowed for the production of superior high quality water with no operator intervention or ion exchange resin replacements in 1700 hours. A comparison between a 100,000 gal/day distillation unit and the 100,000 gal/day EUWP unit may be seen in Table (1).

Table (1) – Comparison of Flash Distillation and Advanced RO Systems

	Distiller	**Advanced RO (EUWP)**
Dimensions (L x W x H)	22 x 10.7 x 9.25	20 x 8 x 8
Weight (lbs)	51,716	33,000
Volumetric Production (gal/day/cu ft)	46	50
Gravimetric Production (gal/day/lb)	1.9	3.7
Operational Availability	< 65%	> 95%
Energy (kWh/1000 gal)	424	65
Cost ($/1000 gal)	$24.75	$3.78

Goals for the Next Generation of Shipboard Desalination Systems

Based on the success of the EUWP program, the Navy selected Shipboard Desalination as a Future Naval Capability Program from FY2010 to FY2014. The primary objective of the program is to develop specifications for advanced shipboard desalination systems that cost the same or less than traditional NSRO systems with 40% less volume, 40% less weight, 65% less energy, greater than 95% operational availability, and 50% lower total ownership costs. The products of the FNC effort will be demonstrator units that will be capable of being installed on a ship. The first two years of the program are dedicated to maturing component level technologies including advanced membrane filtration, antifouling and chlorine resistant RO membranes, energy recovery pumps, and system controls for optimizing shipboard operation. The latter two years of the program are dedicated to designing complete desalination systems (pretreatment and desalination). The completed systems will range in size from 4,000 gal/day to 18,000 gal/day, and 100,000 gal/day.

Design Considerations for Shipboard Desalination Systems

Land based municipal water desalination systems are designed to minimize capital and operational costs and thus energy efficiency is a major concern. For Navy shipboard systems, energy efficiency is an important but not overriding concern since the power to propel a ship through the water dominates overall energy use. Capital costs are important to the Navy, however the Navy will design for a long system lifetime and use highly corrosion resistant materials. Space on a ship is limited and so the desalination system must be compact. High operational availability and minimal maintenance are also required.

There are additional engineering requirements for a successful shipboard design. For combatant craft (e.g., destroyers, cruisers, carriers, submarines), all systems must demonstrate the capability to withstand and continue operating after shock events, to ensure operation after potential explosions near the ship's hull. The systems must also be capable of withstanding vibrations caused by other shipboard equipment. The control systems must meet military requirements for electromagnetic interference, and materials of construction cannot cause issues if a fire exists, limiting plastic components. If the systems are to be installed on existing ships, the construction of the system must be accounted for, i.e., whether it has to be lifted into the ship through the hatchways.

(301c) Regulatory Compliance and Water Treatment Advances: From Research to Application

John Albert, Water Research Foundation, Denver, CO

Water utilities must comply with a myriad of regulations from stage two disinfection by-product rule to the revised total coliform rule. Many of these rules are rigorous and at times conflicting. Precisely navigating the regulatory landscape and meeting water quality objectives while being vigilant about unintended consequences has become a major focus for utilities now more than ever.

Evaluating appropriate technology choices to comply with multiple and/or conflicting water quality goals is a priority for utilities. Currently, a number of decision support tools have been developed to aid utilities in this effort. This talk will highlight one such tool, the Simultaneous Compliance Tool that was recently developed by the Water Research Foundation. The tool is designed for both large and small water systems.

As regulations become more prescriptive, customer expectations increase, and the quality of current source waters is expected to degrade over time many utilities need to consider treatment options above and beyond conventional treatment. Extensive research has been conducted on advanced treatment options such as biological treatment, advanced oxidation processes, membranes, UV, and ozone as a means to meet these challenges.

Moving these technologies from research studies to field application involves extensive benefits analyses including; cost, operations trade-offs, and long-term forecasting for sustainable use. Recent applications of these technologies and case-studies will be highlighted.

See more of this Session: Industry and Government Activities In Water Purification
See more of this Group/Topical: Topical 1: Water Technology for Developed and Developing Countries (see also Separations Division)

Inland Brackish Groundwater as a Resource: Federal Research, Issues, and Challenges

Michelle Chapman
Bureau of Reclamation, Denver, CO
mchapman@usbr.gov

Abstract

The mission of the Bureau of Reclamation is to manage, develop, and protect water and related resources in an environmentally and economically sound manner in the interest of the American public. In the past, that meant building dams and irrigation systems for farmers. Toward the last quarter of the 20st century, Reclamation was called to assist with municipal and industrial water supply through the Title XVI Water recycling and Reuse program for recovering impaired water sources. Now that program has grown into the WaterSMART program (Sustain and Manage America's Resources for Tomorrow) for improving water utility infrastructure, and the Rural Water Supply program to assess potable water supply needs and identify options for fulfilling those needs. Advanced water treatment research has been funded for the past 15 years through the Desalination Act of 1996 and congressional support to the WateReuse Foundation. Through these programs and other federally funded research, we have come to the point where technology is available to treat almost any source of water. The challenge now is to determine the most efficient treatment solution for a particular situation. It is necessary to consider the aquifer water chemistry, productivity, and replenishment, concentrate management, and power options in designing a sustainable customized technical solution to a treatment problem.

Introduction

Why should the Federal Government be concerned with brackish groundwater sources? Congress established the Bureau of Reclamation in 1902 to reclaim the desert of the western United States, providing family farms for our growing country. This was important to solidify our claim on the west. Starting with small irrigation projects, Reclamation negotiated the complex Colorado River Compact between states bordering the Colorado River to allocate its flow. The Compact cleared the way for construction of the Hoover Dam and the All-American Canal starting in 1928. During the Depression, Congress authorized 40 large projects to expand infrastructure and put people to work. Since then Reclamation has evolved to meet new challenges as they arose. It is one thing to design and build a huge dam, but quite another to predict what will happen to fisheries, nesting areas, dam foundations once they are filled with water, irrigated soils, and drainage patterns after the dam and irrigation system are built. Reclamation developed new areas of expertise to solve problems as necessary.

Toward the last quarter of the 20st century, Reclamation was called to assist with municipal and industrial water supply through the Title XVI Water Recycling and Reuse program for recovering impaired water sources. That program has now grown into the WaterSMART program (Sustain and Manage America's Resources for Tomorrow) for improving water utility infrastructure. Congressional support drove several rural water projects in the 1980's and 90's such as the $500 million Mni Wiconi water project. The project is composed of a pumping and filtration plant on the Missouri River and 4,200 miles of pipeline to bring water to central South Dakota for around 10,000 families. Now, however, the Rural Water Supply Act of 2005 gives Reclamation authority to assess, plan, and prioritize these projects rather than having Congress authorize projects individually. As Reclamation delves into the backlog of rural water supply project needs assessments and appraisal studies, the question of treating local brackish ground water arises regularly.

Six of the past ten years have brought significant drought to different parts of the western US. Traditional water supplies are not as reliable as they were historically. Whether one calls it climate change or normal variance in weather patterns, rain and snowfall are not coming when and where they are wanted. Water utility managers realize that they need to have a more diverse portfolio of water sources to draw from. Non-traditional sources, such as municipal wastewater, brackish groundwater, and the sea, are becoming the back-up supply, or in some areas, the primary supply of water.

Fortunately, advanced water treatment research funded for the past 15 years through the Desalination Act of 1996 and through congressional support for the WateReuse Foundation has progressed rapidly. Through these programs and other federally funded research, we have come to the point where technology is available and affordable to treat most any source of water wherever it may be found.

Reclamation's Desalination Research Facilities

Reclamation has two desalination research facilities. The first was the Water Quality Improvement Center (WQIC) in Yuma, Arizona associated with the Yuma Desalting Plant. The original purpose of the WQIC was to evaluate improvements to the desalting process for the larger treatment plant and provide drinking water to the administrative offices at the plant. Changes in pretreatment were tested out in the WQIC. As the treatment process settled, Reclamation offered the WQIC to the public as a place to test treatment ideas for Colorado River water. Clients could choose from a variety of water sources: irrigation return flow from the canal or river water; raw or pretreated; concentrate or permeate; chlorinated or not. There are a variety of reverse osmosis skids, micro-ultra filtration, chemical injection, and conventional pretreatment systems available for clients to use in their studies.

The second facility is the Brackish Groundwater National Desalination Research Facility (BGNDRF) in Alamogordo, NM. Built in 2005 through the Desalination and Water Purification Research Program, the facility is designed to test and evaluate treatment processes and concentrate management solutions. There are four wells with total dissolved solids levels ranging from 550 to 4200 mg/L. There are four outside test pads for medium (60 gpm) to large (130 gpm) pilot studies and six bays inside for test systems treating 30 gpm or less. The facility has two evaporation ponds for concentrate management studies, an agricultural research area and space for wind and solar powered process development.

Projects tested at the facility so far address many aspects of brackish aquifer utilization. There are projects for testing alternatives for desalination concentrate minimization or elimination, new membrane development, processes for produced water treatment, a new method of electrodialysis, and a variety of brackish groundwater pretreatment processes. The number of projects is increasing each year. There are at least eight scheduled for 2012.

Issues to Resolve

We really have a lot to learn about brackish groundwater aquifers. Traditionally, the USGS and EPA have gathered data on fresh water wells. Since people generally do not use saline wells, data is not collected. The oil and gas industry have information on deep saline aquifers near oil and gas wells, but their databases are not available to the public. What happens to brackish aquifers as water is withdrawn? Can brackish aquifers be used in a sustainable manner? There are many sides to the issue – or many issues depending on ones perspective.

- Engineering perspective
    - How can we make BW desalination more robust?

- How can we integrate alternative energy sources into the desalination process?
    - Can we optimize the mineral recovery process to be as reliable as food processing?
    - What is the best design point – lowest capital cost or most reliable?
- Hydro-geologist perspective
    - How is saline groundwater connected to fresh groundwater and surface water?
    - How are these other water sources affected by extracting saline groundwater?
    - Will the quality of the saline groundwater change over time?
    - Can we sustainably manage the concentrate?
- Legal perspective
    - Who has rights to brackish water?
    - Are those who desalt brackish groundwater responsible for the concentrate until it reaches the ocean?
    - Are they responsible to ensure that it is properly mixed when it enters the ocean?
- Socio-economist perspective
    - Is desalted brackish water more precious than fresh because of the added cost?
    - Should it be used to wash the car and water cattle?
    - Is it for industry or only for drinking?
    - If it is used for industry, should the industry pay back the development cost?
    - Should it be reused often or returned to the hydrologic cycle?
    - Should the government support desalination projects in the desert or encourage migration to water rich areas.

Conclusion

As we go forward in our efforts to make rural Western US a kinder place by providing water for all no matter where they live, we need to remember our past mistakes in reclaiming the west, study potential consequences, and find answers to these questions to aid us in our plans.

References

Brief History Bureau of Reclamation - http://www.usbr.gov/history/2011NEWBRIEFHISTORYV1.pdf

2008, Brokaw, C. "Mni Wiconi water reaching Pine Ridge reservation" News From Indian Country, accessed from http://indiancountrynews.net/index.php?option=com_content&task=view&id=4387

Mni Wiconi Rural Water Supply Project (15.522) Agency: Department of the Interior Office: Bureau of Reclamation. Accessed from: http://www.federalgrantswire.com/mni-wiconi-rural-water-supply-project.html

Yuma Water Quality Improvement Center: http://www.usbr.gov/lc/yuma/facilities/wqic/yao_facilities_wqic.html

Brackish Groundwater national Desalination Research Facility Fact Sheet: http://www.usbr.gov/pmts/water/research/BGNDRF%20FACT%20SHEET%20Final.pdf

(301e) Electrodialysis Reversal Pilot Studies Performed At the Brackish Ground Water National Desalination Research Facility (BGNDRF) In Alamogordo, NM

Neil E. Moe[1], John Barber[1], Ali Sharbat[2] and Abbas Ghassemi[2], (1)GE Water & Process Technogies, Las Cruces, NM, (2)Institute for Energy and the Environment, New Mexico State University, Las Cruces, NM

Electrodialysis Reversal (EDR) technology uses voltage driving force and ion selective membranes to remove salt from water. The "Reversal" in EDR refers to the process of periodically reversing the polarity of the electrodes in order to inhibit scale formation. EDR has been an established desalination technology for almost 40 years, and yet the understanding of EDR is less developed than that of other technologies such as distillation and reverse osmosis. For example, all the major suppliers of reverse osmosis membranes also provide advanced simulation tools that enable engineers to reliably design systems. In contrast, there is no such design software publicly available for EDR and very few literature references that provide operating data for commercial sized systems. EDR is cloaked in mystery.

Extensive studies were performed using full sized pilot EDR equipment, where the electrodes, spacers, and membranes were sized identical to commercial systems, while the number of cell pairs was typically 40, compared to 600 at most commercial installations. The experiments were performed at BGNDRF utilizing the various sources of brackish water available on site. Experimental variables included water chemistry, temperature, recovery, velocity, volts/amps, stack configuration, and membranes. Thousands of individual experiments were performed and over 900 water samples were collected for analysis.

Results were analyzed in a number of ways. First, the relationship between volts, amps, and cut rate of dissolved salts is defined as function of operating conditions. Second, the selectivity for various ions was determined, indicating that selectivity was highest at high temperatures and low cut rates. Third, limiting current was determined and found to have unexpected functional dependencies. Temperature had very little effect over the range 7 – 40 C, the nature of the ionic species had no discernible effect, and limiting current was nearly linearly dependent on fluid velocity. The quantitative coupling of operating conditions (temperature and flow velocity) with performance (volts, amps, conductivity) and water chemistry led to greatly increased understanding of EDR process and provided the foundation for an EDR process model.

See more of this Session: Industry and Government Activities In Water Purification
See more of this Group/Topical: Topical 1: Water Technology for Developed and Developing Countries (see also Separations Division)

Feasibility Study of High Cut Rate of Dissolved Mineral Salts in Electrodialysis Reversal (EDR) Process

Connor Hanrahan[1], Ali Sharbat[1], Neil E. Moe[2], John Barber[2] and Abbas Ghassemi[1]
(1) Institute for Energy and the Environment, New Mexico State University, Las Cruces, NM; (2) GE Water and Process Technologies, Las Cruces, NM

Abstract

Relatively high capital expenditure (CapEx) of desalination technologies has limited the practical employment of saline and brackish water desalination. Hence, any attempt to reduce CapEx will result in greater utilization of desalination technologies. Conventional separation percentages of total dissolved salts in EDR technologies is between 40-60% per stage for current industrial-scale systems [1,2]. Thus, depending on the saline/brackish water quality, most EDR systems are designed to desalinate water in multiple stages, which results in higher CapEx costs. In this research, the feasibility of increasing the cut rate of dissolved mineral salts in an EDR process has been studied in an effort to lower CapEx costs by reducing the number of required stages.

Full pilot-scale experimentation was performed on a 41 cell pair electrodialyser with 12 gpm influent flow rate at the Brackish Groundwater National Desalination Research Facility (BGNDRF) in Alamogordo, New Mexico. Brackish feed-water was provided from a deep aquifer in the Tularosa Basin with an average conductivity of 1794 μS/cm, which is within the spectrum for water quality in the southwestern region of the United States[3,4]. Using this feed water, short term limiting current experiments were performed that demonstrated an entitlement of 88-99% cut rate of dissolved solids per stage, directly related to the limiting current of the system. Next, long-term experiments were performed to demonstrate the feasibility of continuous high-removal operation. More than 150 hours of continuous operation were successfully completed at ≥ 90% removal of dissolved solids.

Introduction

According to the 2003 "World Water Development Report" by the United Nations, more than 50 percent of nations will face water stress or water shortages by 2025, and as much as 75 percent of the world's population could face water scarcity by 2050. Fueled by rapid population growth, much of the Southwest United States is already witnessing an increased demand on their limited fresh water supply, and is looking towards its extensive brackish water resources to address this issue. Where inland brackish water desalination has the potential to create "new" water resources in these arid regions, implementation is hindered by the relatively high cost of treating brackish waters.

Multiple processes exist for treating saline waters such as reverse osmosis (RO), electrodialysis reversal (EDR), evaporation, membrane distillation and ion exchange. Where each technology supports a different role in treating saline waters, electrodialysis reversal (EDR) has proven particularly useful for treating hard waters. EDR utilizes an electric field to remove charged particles from a saline solution through permselective ion-exchange membranes. Where sparingly soluble salts accumulate on the surface of membranes during standard desalination procedures, EDR has the advantage of a periodic polarity reversal in which the salts are cleared from the membrane surfaces before becoming permanently attached. This makes EDR particularly well-suited for treating brackish waters where the concentrations of sparing soluble salts, such as calcium sulfate and calcium carbonate, are high.

However, EDR suffers from relatively high CapEx costs, when compared with an RO process, in which the stack components are the primary cost. In fact, EDR is most competitive when a one or two stage system is used to desalt raw water sources [5]. Though extensive laboratory testing has shown an entitlement of up to 99% removal per stage for EDR systems, current EDR plants are operated at 40-60 % removal per stage [1,2]. Therefore, an opportunity exists to reduce the CapEx costs of EDR by increasing the single-stage removal and decreasing the number of required stages.

Full pilot-scale experimentation at BGNDRF has successfully completed more than 150 hours of continuous operation at >90% single-stage removal, with 12 gpm influent flow from a brackish water aquifer in the Tularosa Basin near Alamogordo, New Mexico.

Theory

Significant modeling work has been done to show that as dissolved salts are passed across permselective membranes, the electrolyte concentration decreases in the boundary layer, which is formed on the desalting surface of the membrane [6]. At the limiting current density, the concentration of dissolved salts on the surface of the ion-exchange membranes approaches zero, where the transmembrane flux does not increase with increasing driving force [7]. Above the limiting current density, water molecules will dissociate and lead to the precipitation of scale-forming inorganics or cause membrane damage. Care must therefore be taken to understand the limiting current density of a particular EDR assembly with its specified feed water.

A current-voltage curve is typically used to determine the membrane resistance, the limiting current and the boundary layer thickness of a system containing ion exchange membranes [8]. The curve is used to relate the current passed through a membrane with the voltage drop across the membrane and its adjacent boundary layer. By adjusting the electrical potential between the electrodes and monitoring the change in current, a resistance profile was used to determine the limiting current (I_{lim}) from the inflection of the V/I vs 1/I plot (Fig. 1).

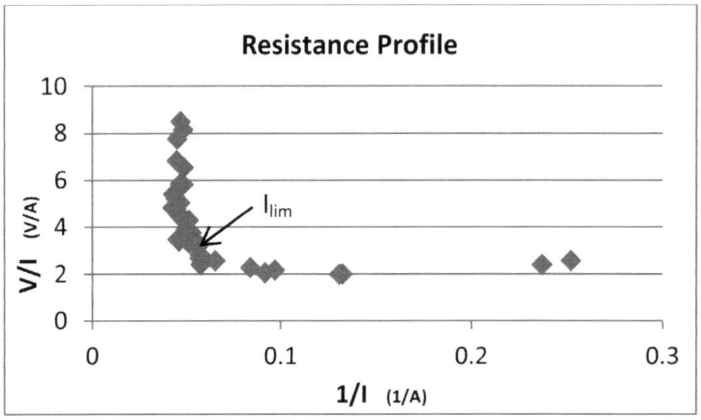

Fig. 1 V/I vs. 1/I plot for determining limiting current.

The current-voltage curve is the most common graphical technique for determining the limiting current of an EDR system. The current-voltage curve from our data determined a limiting current density of 18.72 A, which stands as the theoretical limit of operation in order to avoid water splitting and the subsequent membrane damage/scaling. However, experimental data on '% Removal vs. Current' reveals that the stack is capable of operating above 20 amps without witnessing the drastic pH changes caused by water splitting.

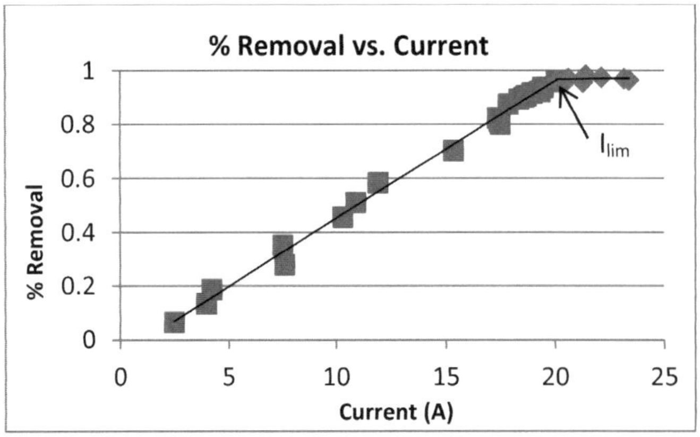

Fig. 2 '% Removal vs. Current' plot for determining limiting current.

This discrepancy of results in determining the limiting current density is most likely due to the effects of very high, single-stage removal. Where most researchers performing limiting current experiments are working with higher salinity feed-water and laboratory-scale systems, by the time our stack removes 90% of the dissolved minerals, the diluate stream begins to add to the total stack resistance, due to a lack of ions. The current-voltage curve proposed by other researchers [6-8] addresses the resistance across the membrane surfaces and adjacent boundary layers, but not the influence of resistance caused by a nearly demineralized diluate stream. Based on this observation, our experimental design was based upon the '% Removal vs. Current' plot, targeting 19.72 amps (97% of I_{lim}).

Materials and Methods

Full-scale pilot studies were carried out at BGNDRF, using brackish water from a deep aquifer in the Tularosa Basin near Alamogordo, New Mexico. The system includes a multi-media filtration tank, a cartridge filter, an EDR stack, pumps, control valves and a main control panel. The raw feed water was passed through a multi-media filter and a GE cartridge filter before reaching the hydraulic system, where it was separated into concentrate and diluate feed to the EDR stack.

The experiments were performed using an Ionics Aquamite I EDR stack (23x25.5x50 cm^3) with two electrical stages of 41 cell pairs each. GE Cation-CR67-HMR and Anion-AR204-SZRA membranes were used, separated by Mark IV spacers (0.76 mm thick), with an effective membrane surface area of 3,450 cm^2. Due to experimental design considerations, only the first electrical stage was used in the removal of dissolved solids, while the voltage of the second stage was set only to deter reverse electrodialysis (no current flow). Under these design parameters, the system had a total effective membrane area of 14.15 m^2 and an average current density of 59.4 A/m^2 (aprox. 97% of limiting current density). The voltage between the electrodes was controlled by a power supply that provided a constant electrical potential difference. An oscilloscope was used to measure the voltage and current between the electrodes (stage V & stage A), and a multimeter was used to monitor the stack voltage for irregularities. The polarity reversal period was set for 15 min.

Due to peculiarities of the system, the positive and negative cycles operate with slightly different product flow rates – 7.5 and 7.8 gpm respectively. As the system provides a constant voltage, this difference in flow rate causes a variance in amperage, which was remedied by targeting 90% removal based on separate drive setting for the positive and negative cycles. The system was therefore operated in positive polarity at 66.2 V and around 19.2 A, and in negative polarity at 71.3 V and around 21.3 A, which led to minor discrepancies in product and concentrate water conductivities.

Due to the nature of the experiment, and the high potential for $CaCO_3$ scaling from the feed water, a continuous acid feed of 15% HCl was fed to the concentrate stream to keep the LSI below 2.1, and an ECIP solution of pH 2.5 was run continuously to the electrode chambers.

A limiting current experiment was performed at the beginning of the test to ensure that the stack performance was optimal and to provide a target range for 90% single-stage removal.

Results and Discussion

Table 1 –
Run-Time Data at 90% Removal

Polarity	Conductivity (µS/cm)			pH			Voltage (V)	Current (A)	Stack Voltage (V)	% Removal
	Feed	Product	Concentrate	Feed	Product	Concentrate				
Positive	1794	159.5	5616	8.3	8.2	8.1	66.2	19.1	62.1	91.1
Negative	1794	197	5930	8.3	8.1	8.1	71.3	21.3	67.8	89.0

Table 2 –
Water Quality Data at 90% Removal

	Ca^{2+}	Mg^{2+}	Na^+	K^+	Sr^{2+}	Cl^-	SO_4^{2-}	HCO_3^-	Fl^{2+}	Si	PO_4^{3-}
FEED	58.5	13.5	326.6	5.95	1.9	38.6	632.3	196	2.61	26	0.6
Product	6.5	1.7	28	<0.5	.22	1.9	50	34.5	0.5	29	< 0.4
Concentrate	278.2	56.862	842	22	8.7	123	2105	251.7	4.5	26.5	0.7

* Concentrations, mg/L

The average run-time data and water quality analyses are presented in tables 1 and 2. The feedwater conductivity varied between 1753 and 1849 µS/cm, and the product and concentrate conductivities varied according to polarity reversal period as discussed previously. The electrical potential and current remained constant throughout operation. The stack voltage was monitored to ensure that the stack was operating properly, and it also remained constant throughout the experiment. The product blowdown was 7.5 gpm in positive polarity and 7.8 gpm in negative polarity, while the concentrate blowdown remained constant at 2.2 gpm.

Conclusion

In an effort to reduce the required CapEx for EDR processes, high single-stage removal was demonstrated using brackish groundwater from a deep well in the Tularosa Basin, near Alamogordo, New Mexico. Based on previous limiting current experiments, more than 150 hours of continuous operation were successfully completed targeting 90% single-stage removal.

Acknowledgements

We wish to thank Dr. Abbas Ghassemi, Reclamation, General Electric, and entire staff at BGNDRF for their continued support.

References

1. Veza J.M. (2001) "Desalination in Canary Islands: An update." Desalination 133
2. Valero, F., Arbós, R. (2010) "Desalination of brackish river water using Electrodialysis Reversal (EDR): Control of the THMs formation in the Barcelona (NE Spain) area." Desalination, 253.
3. Gleick, P. (2010) "Roadmap for sustainable water resources in southwestern North America." PNAS, 14 Dec
4. Lansford, R., Hernandez, J., Enis, P.J., Truby, D. and Mapel, C. (1987) "Evaluation of available saline water resources in New Mexico for the production of microalgae." FY 1986 Aquatic Species Program Annual Report; Solar Energy Research Institute, Golden, Colorado, SERI/SP-231-3071.
5. Reahl, E. R. (2008). "Half A Century of Desalination with Electrodialysis." from http://www.gewater.com/pdf/Technical Papers_Cust/Americas/English/TP1038EN.pdf.
6. Tanaka, Y., Iwahashi, M., Kogure, M. (1994) "Distribution of electrodialytic condition in an ion-exchange membrane electrodialyzer and limiting current density." Journal of Membrane Science. 92
7. Ibanez, R., Stamatialis, D.F., Wessling, M. (2004) "Role of membrane surface in concentration polarization at cation exchange membranes." Journal of Membrane Science, 239.
8. Długołęcki, P., Anet, B., Metz, S., Nijmeijer, K., Wessling, M. (2010) "Transport limitations in ion exchange membranes at low salt concentrations." Journal of Membrane Science, Vol. 346, No. 1

TREATMENT OF AMMUNITION WASTEWATER USING ULTRASONIC FENTON (US-FENTON) PROCESS

YangangLi[a, b], Wen-Pin Hsieh[a], RovshanMahmudov[a], and C.P. Huang[a]*

[a] Department of Civil and Environmental Engineering, University of Delaware, Newark, DE 19716, USA

[b] College of Water Resources and Architecture Engineering, Northwest A&F University, Yangling, Shaanxi, 712100,China

[*] Corresponding author. C. P. Huang (Fax: 1-302-831-8428; Email: huang@udel.edu)

Abstract

Ammunition wastewater contains several persistent compounds, such as 2,4,6 trinitrotoluene (TNT), 2,4 dinitroanisole (DNAN), and 3-nitro-1,2,4-triazol-5-one (NTO), that are recalcitrant during conventional treatment. In this study, the wastewater collected from a regional ammunition facility was screen treated with various methods, including UV-photolysis, Fenton, photo-Fenton and ultrasound-Fenton (US-Fenton). Results showed that US-Fenton was the best method among all screened and was selected for further study. The effect of pH, temperature, reaction time, US energy intensity, initial TOC concentration, and the molar ratio of iron to hydrogen peroxide were investigated. TOC, COD, and color removal increased with decreasing pH. TOC, COD, and color removal were increased with increase in temperature and US intensity. Color removal was very fast, reaching 80 % in 10 minutes. Compared to color removal, TOC and COD removal were slower, only about 20 % in 10 minutes, then approached 65 % and 93 % in 120 minutes, respectively. The optimal molar ratio of iron to hydrogen peroxide was 500. The carbon oxidation numbers were calculated for different treatment options. The results showed that the carbon oxidation number increased with increase in time. The toxicity of treated wastewater was evaluated by monitoring the respiration rate of *E. coli* exposed to the ammunition wastewater before and after treatment. Results indicated that sample treated for 300 minutes improved the respiration rate of *E. coli* compared to untreated samples.

Keyword: Ammunition wastewater, Ultrasound, Fenton, Total organic carbon (TOC), Chemical oxygen demand (COD), Toxicity.

1. Introduction

Ammunition wastewater from the wash-down of energetic materials handling equipment at military facilities has a complex chemical and physical characteristics and contains a number of harmful contaminants, such as metals, nitroaromatic compounds, and various oxyanions that require proper treatment before discharge [1, 2]. Nitroaromatic compounds have been widely used in the manufacture of explosives since World War II [3, 4]. Some metals and nitroaromatic compounds are toxic to humans and animals, and are generally very persistent in the environment if discharged without proper treatment. In addition, environmental transformation products of nitroaromatic compounds (i.e. azoxy- and azo-compounds) can be even more toxic than the parent nitroaromatics [5-7] .

Much research has been conducted in order to develop an effective ammunition wastewater treatment technology that would be applicable for wide changes in composition of these wastewaters. Several biological and physical-chemical treatment processes have been used. Yin et al. studied reductive transformation of TNT using aerobically grown *Escherichia coli* cultures [8]. Kang et al. investigated abiotic transformation of TNT via oxidative-coupling reaction using Mn oxide [9]. Oh et al. studied the removal of TNT and RDX using Fenton process as a subsequent treatment following iron pretreatment.[10]. Yardin et al. investigated the remediation of TNT contaminated soil using cyclodextrins coupled with Photo–Fenton treatment [11].

Advanced oxidation process (AOP) has been shown as promising to treat wastewaters containing recalcitrant contaminants. Particularly, AOP based on ultrasound irradiation has been found highly effective. Ultrasound manifests its physical and chemical effects through the phenomenon of cavitation, which can be defined as the cycle of formation, growth, and collapse of micro bubbles within the media. Transient collapse of cavitation bubbles results in concentration of energy and generation of extremely high local temperature and pressure. Temperature as high as 5000 ^{o}K and pressures up to 500 bar have been suggested during the transient collapse of the micro bubbles [12]. Under these conditions, entrapped molecules of dissolved gases (e.g. vaporized water and solutes) can be brought to an excited state and undergo hemolytic bond breakage to generate radicals; when water is the solvent, $OH^{\bullet}$, $H^{\bullet}$, and $HO_2^{\bullet}$ radicals are generated [13]. Hydroxyl radical is among the strongest oxidants with an E_H^o value of 2.33V and can react non-selectively with a host of organic and inorganic compounds [14]. Free radicals generated during water and oxygen dissociation combined with local high temperatures and pressures from formation and collapse of micro bubbles induce/accelerate oxidation reactions in the liquid medium [15].

$$H_2O \text{ ---))))} \rightarrow H^{\bullet} + OH^{\bullet} \quad (1)$$
$$O_2 \text{ ---))))} \rightarrow 2O \quad (2)$$
$$2O + H_2 \rightarrow 2OH^{\bullet} \quad (3)$$
$$H^{\bullet} + O_2 \rightarrow OH^{\bullet} + O \quad (4)$$

However, total mineralization of organic pollutants by means of ultrasound irradiation alone is very energy consuming, and in many cases has low efficiency, therefore its application for a full scale industrial plant is still impractical. To overcome these limitations, various combined ultrasound treatment systems have been proposed [16]. These treatment methods combine high energy confined within micro bubbles generated by ultrasound with chemical sources of hydroxyl radical formation. Over the past several years, there has been an increasing interest in the sonochemical process for the treatment of recalcitrant contaminants in aqueous solutions that are hardly amenable to conventional treatment [17-23]. Sonochemical treatment includes ultrasound treatment coupled with oxidants such as ozone [13, 14, 24], hydrogen peroxide (H_2O_2)[25, 26] , electrochemical methods [27, 28], Fenton's reagent[23, 29] , photocatalysts[29], and other catalytic oxidation processes [30]. Particularly, the combination of ultrasound with Fenton's reagent has evolved as a promising AOP for the treatment of ammunition wastewater. In Fenton oxidation process, the reaction of Fe^{2+} with H_2O_2 generates $OH^{\bullet}$ radicals as shown in Eq. 5.

$$H_2O_2 + Fe^{2+} \rightarrow Fe^{3+} + OH^{-} + OH^{\bullet} \quad (5)$$
$$RH + OH^{\bullet} \rightarrow H_2O + R^{\bullet} \quad (6)$$

Hydroxyl radicals react with organic compounds to generate other radicals as shown in Eq. 6; organic radicals are unstable and degrade rapidly. The results of several studies have shown that combined ultrasound and Fenton process were better than ultrasound or Fenton process alone.

In this study, ultrasound (US) assisted Fenton reaction was applied for the treatment of ammunition wastewater. The objectives of this study were (1) to determine the optimal conditions for the treatment of ammunition wastewater, and (2) to determine the effect of treatment on the short-term toxicity of ammunition wastewater.

2. Experiments

2.1 Materials

All chemicals used in this study were ACS (American Chemical Society) certified grade and obtained from either Fisher Scientific or Sigma Aldrich and used without further purification.

The ammunition wastewater samples generated from wash-down of the equipment handling energetic materials and occasional sump out operations was collected for use in this study. Table 1 shows the physical-chemical characteristics of the wastewater.

2.2 Analytical methods

Chemical oxygen demand (COD) was measured by the Hach colorimetric method at 600 nm using a Hach spectrophotometer (Hach DR/2000) according to The Standard Methods for the Examination of Water and Wastewater (1990). Residual hydrogen peroxide concentration in treated ammunition wastewater was analyzed using the titanium sulfate method [31]. Hydrogen peroxide interferes with COD measurements, and therefore the final COD values of treated ammunition wastewater were corrected for the residual hydrogen peroxide.

Total organic carbon (TOC) was measured with a TOC analyzer (Apollo 9000HS, Teledyne-Tekmar). Soluble ferrous iron concentration [Fe^{2+}] was determined using the 1,10-phenanthroline colorimetric method. To avoid the interference of ferric iron, ammonium fluoride was added as the masking agent. Total soluble iron concentration was determined by reducing ferric to ferrous ion by hydroxylamine under acidic conditions. Total iron concentration then was measured by the same method as Fe^{2+}concentration. Ferric ion concentration was estimated by subtracting ferrous ion from total iron.

The concentration of anions was analyzed using an ion chromatograph Dionex 500 equipped with a conductivity detector, AG-11 guard column, and AS-11 metal-free anion column (25 mm×4 mm i.d.). The eluent was a mixture of 87% deionized water, 10% 0.2 M NaOH and 3% acetonitrile. The effluent flow rate was 1 mL/min and the injection loop volume was 50 μL. Cations were determined using Inductively Coupled Plasma Spectrometer (ICP-OES) (Optima 7300 DV, Perkin-Elmer Corp). The mixed liquor and all other samples were digested with nitric acid before ICP-OES analysis for the determination of total metal concentration.

The toxicity of treated and untreated wastewater was determined with short-term toxicity test using *Escherichia coli*. The *E. coli* culture medium was prepared by dissolving 1.6 g KH_2PO_4, 1.6 g K_2HPO_4, 1.0 g NaCl, 2.5 g glucose, 4.0 g $(NH_4)_2SO_4$, 0.7 $MgSO_4$ and 0.5 g sodium citrate in 1 L of distilled water. The finial pH was adjusted to 7.0 using 4 M solution of NaOH. Culture medium was sterilized in an autoclave at 125°C for 15 min. Then 50 mLof sterilized culture medium was carefully transferred to 125 mL Erlenmeyer flask and 0.5 mL of *E. coil* seed solution was added. The flasks were placed on a reciprocal shaker and *E. coli* culture was allowed to grow at a constant temperature of 37°C for 12 h. To test the effect of the chemical treatment on the toxicity of ammunition wastewater, one mL of the following solutions were added to flasks containing *E. coli* cultures: original wastewater, original wastewater diluted 10 times, and wastewater treated for 60 and 300 minutes, respectively. After predetermined time intervals, one-mL aliquots of liquid culture medium were withdrawn and analyzed for carbon dioxide, according to the method proposed by Jardim et al. [32]. The flasks were shaken on a reciprocal shaker throughout the entire process.

2.3 Sonochemical reactor

Figure 1 shows the schematic diagram of the experimental apparatus. Sonochemical experiments were carried out using the ultrasoound generator (Cole-Parmer 600-Watt, 20 kHz,) equipped with a titanium probe transducer (Cole Parmer, Model CV 17). The experiments were conducted in a 40-mL borosilicate glass vial. The reactor was filled with 30 mL of ammunition wastewater and 5 mL of Fenton reagent. To maintain constant ionic strength, 0.5 mL of 0.7 M of stock NaCl solution was added to yield constant ionic strength of 10^{-2} M. The reactor was sealed with a Teflon-lined rubber stopper and immersed into a jacketed glass reactor containing DI water, in which the temperature was maintained at 25°C by a water circulation and temperature control system (Frigomix 1495, Fisher Scientific).

3. Results and Discussion

3.1 Screening of various treatment methods

Several methods were screened for treating the ammunition wastewater, including, UV-photolysis (UV $_{254\ nm}$), Fenton, photo-Fenton, and ultrasound-Fenton using TOC as the main performance indicator. The initial TOC concentration of the ammunition wastewater was 710±50 mg/L and pH of 2.8±0.2. The TOC concentration of the ammunition wastewater was monitored for 1 h and the results are shown in Figure 2. Results showed that the degradation of TOC was relatively low, less than 10%, when the ammunition wastewater was treated with only UV photolysis, $FeSO_4$, or H_2O_2. Low TOC degradation was also observed for Fenton and photo-Fenton processes. The highest TOC removal at 58% was observed when the wastewater was treated by a combination of ultrasound and Fenton regent with the concentration of [Fe^{2+}] and [H_2O_2] of 5×10^{-4} M and 0.25 M, respectively, and US intensity of 300 watts. The oxidation efficiency followed the order: ultrasound + Fenton > Fenton > photo+Fenton > H_2O_2 > $FeSO_4$ > UV. The ammunition wastewater had strong color that prevented the UV light from passing through the sample, which could affect the treatment efficiency of the UV and the photo-Fenton processes compared to US-Fenton process. The results clearly indicated the superiority of the ultrasound-Fenton process over other treatment processes studied. Therefore further studies ware focused on the ultrasound-Fenton method.

3.2 Effect of ultrasound intensity

The sonicator used in this research had a constant output frequency of 20 kHz. Higher power intensity results in the formation of cavities of higher temperature [33]. Figure 3 shows the degradation of color (measured as absorbance at 348 nm), TOC and COD as a function of power intensity. The color removal increased slightly from 83.7 to 87.7 % when the power intensity was increased from 0 to 300 watts. The removal of TOC and COD, however, improved significantly when the power density was increased. TOC and COD removal increased from 30.5 and 13.6 % to 73.6 and 55.7%, respectively, when the power intensity increased from 0 to 300 watts. Higher degradation percentage of ammunition wastewater was achieved at higher power intensity. Increase in sonic energy increases the number of generated cavities and the cumulative pressure [33]. This results in a greater $OH^\bullet$ radical generation rate which will enhance the decomposition of organic compounds. On the other hand, collapse of cavities enhances the reaction of OH• radicals with organic compounds by greatly accelerating the mass transfer rates.

3.3 Effect of Hydrogen Peroxide to Ferrous Iron molar Ratio

Figure 4 shows the removals of COD, TOC, and color at fixed ferrous iron dose of 5×10^{-4} mole/L and varying the concentration of hydrogen peroxide to yield various molar ratios of hydrogen peroxide to ferrous iron from 10 to 1500. As the molar ratio was less than 500, the removal of TOC and COD increased with the molar ratio of H_2O_2/Fe^{2+}. The removal of TOC deteriorated when the molar ratio was increased from 500 to 1000. COD removal increased with the molar ratio of H_2O_2/Fe^{2+}. The maximum removal of TOC and COD at 69 and 54%, respectively, was achieved at H_2O_2/Fe^{2+} molar ratio of 500 and 1500. In Fenton oxidation process, the reaction of Fe^{2+} with H_2O_2 generates OH• radicals which are strong oxidants (Eq. 5). Reaction between hydroxyl radical and organic compounds produces organic radicals as shown in En. 6. The organic radicals are unstable and decompose rapidly. At low H_2O_2/Fe^{2+} molar ratios there is insufficient supply of hydrogen peroxide for the generation of hydroxyl radical (Eq. 1 and Eq. 2). Therefore increase in H_2O_2/Fe^{2+} molar ration will increase $OH^\bullet$ generation and lead to better TOC removal. On the other hand, if the H_2O_2/Fe^{2+} molar ratio is too high, one or more side reactions would occur as shown in Eqs. 7 -10, i.e.,

$$H_2O_2+Fe^{3+}\rightarrow Fe^{2+}+HO_2^\bullet+H^+ \quad (7)$$

$$H_2O_2+OH^\bullet\rightarrow HO_2^\bullet+H_2O \quad (8)$$

$$Fe^{3+}+HO_2^\bullet\rightarrow Fe^{2+}+O_2+H^+ \quad (9)$$

$$Fe^{2+}+HO_2^\bullet\rightarrow Fe^{3+}+HO_2^- \quad (10)$$

The above reactions result in the consumption of hydrogen peroxide as well as hydroxyl radical which impedes the degradation of organic compounds in question. It is interesting to note that color removal was independent on the molar ratio of hydrogen peroxide to ferrous ion.

3.4 Effect of pH

Fig. 5 shows the effect of the initial pH on the treatment of ammunition wastewater by US-Fenton process. TOC, COD, and color removal were significantly affected by pH. The removal was greater at low initial pH; at pH 2.0, TOC, COD and color removal were 64, 66, and 74%, respectively. In contrast, at pH 10.0, the degradation of TOC, CO and color were 48, 12, and 12%, respectively. Other studies have showed similar pH effect. Zhang et al. (2009) examined the effect of initial pH on the discoloration of C.I. Acid Orange 7 in the pH range of 2 to 5 and reported that the discoloration increased with decrease in pH and reached the maximum value at pH of 2. Our results agreed with those reported by Hermosilla et al. (2009) and Zhang et al. (2009) who studied the effect of pH on the treatment of landfill leachate in the pH range between 2.0 and 4.0 [34]. The optimal pH was found to be at 2.5. It is known that both Fe^{2+} and Fe^{3+} will precipitate into hydroxides at alkaline pH, which in turn will lead to decrease in the removal efficiency at high pH values. Iron precipitation at higher pH values could be visually observed by the deposition of brown precipitates at the bottom of the reactor. Moreover, the oxidation potential of hydroxyl radicals declined at higher pH values

3.5 Effect of Temperature

The effect of temperature on the treatment of ammunition wastewater was investigated and the results are shown in Figure 6(a). When the temperature was raised from 25 to 40°C, the removal of TOC, COD, and color increased and reached 76%, 87% and 93%, respectively. The highest increase was observed for COD removal and increase in carbon oxidation state.

In order to describe the relationship between of the organic matter decomposition rate and temperature, COD removal rate was used [35]. It was found that the COD removal rate followed first order rate equation as given by Eqn. (7):

$$-\frac{d[C]}{dt} = \kappa[C] \text{ or } C = C_0 \exp(-\kappa t) \qquad (7)$$

Where: C = the concentration of COD at time t, C_0 = initial concentration, κ = removal rate constant and t = reaction time. Figure 6(b) shows the effect of temperature on the first order rate constant for the COD removal. The rate of COD removal increased sharply when the temperature increased from 20 to 35°C followed by a gradual increase as the temperature increased from 35 to 60°C. Such behavior can be attributed to the fact that the rate of redox reactions is generally accelerated by increasing the temperature. At higher temperatures, the reaction of hydrogen peroxide and iron (Fe^{2+}/Fe^{3+}) proceeds faster, this can increase the rate of hydroxyl radical generation and consequently the degradation of organic compound[33]. From the degradation mechanism of US, it is known that when the cavitation bubbles explosively collapse, the temperature and pressure in the bubbles can reach up to several thousand Kelvin and several hundred atmospheres, respectively. Under these conditions, organic compounds decompose directly by pyrolysis inside the bubbles [36]. However, increase of temperature over a certain level may result in an adverse effect. In the US process, higher temperatures are likely to facilitate bubble formation due to an increase in the equilibrium vapor pressure; however, at elevated temperatures bubbles contain more water vapor that may cushion bubble implosion and consequently reduce the maximum temperature obtained during bubble collapse [37].

3.6 Effect of Reaction Time

Figure 7(a) shows TOC, COD, and color removal as a function of reaction time. Color removal was very fast and within the first 10 minutes the degradation reached 85%. In contrast within the first 10 minutes TOC and COD removals were significantly slower than color removal, at only 19 and 21%, respectively. However, after 10 minutes color removal exhibited no obvious change up to 300 minutes of reaction time reaching the maximum removal of 87%. TOC and COD removals increased gradually reaching 55% and 92%, respectively, after 120 minutes of the reaction time, after which the TOC and COD removal rates slowed down significantly reaching 60% and 95%, respectively, in 300 minutes. The removal efficiency of TOC was lower than that of COD. In order to explain better COD removal, the average carbon oxidation state was determined. The average oxidation state is defined as follows:

$$\text{Average oxidation state} = 4[1-(\text{COD/TOC})] \qquad (11)$$

Where: COD is expressed in moles of O_2 / L and TOC is in moles of C / L. Although the hydroxyl radicals and cavitations created by Fenton reagent and US have strong oxidability, they cannot oxidize all organic carbon into inorganic carbon, such as carbonate, bicarbonate or carbon dioxide [34]. The main reasons are the complicated composition of the ammunition wastewater and a complex Fenton oxidation mechanism. Therefore, only a portion of the organic carbon is mineralized into inorganic carbon. In addition, the decomposition of higher molecular weight organic matter generates lower molecular weight organic intermediates, which can lead to increase of the average oxidation state in the residual sample solution, but not the COD value. Increase in the average oxidation state means that COD to TOC ratio should decrease, just like Figure 7(a) shows, with time increasing from 0 to 300 minutes, the average oxidation state went up from -2.1 to 3.3.

Residual concentrations of ferrous iron, ferric iron, and hydrogen peroxide were measured during the reaction as well (Figure 7(b)). During the first 30 minutes, most of ferrous iron was consumed and its concentration decrease from 28 mg/L to 4.1 mg/L. At the same time, ferric iron was produced rapidly. The disappearance of Fe^{2+} and generation of Fe^{3+} occurs according to Eqn.(1) and (6). In the beginning of reaction, ferric iron rapidly reacted with H_2O_2 to generate HO_2·and ferric iron. It is also possible that some amount of ferrous iron was consumed by a different side reaction because ferrous iron consumption rate was greater than the ferric iron generation rate. As the reaction continued, the concentration of ferrous iron in the solution increased slightly to 4.8 mg/L at the 60 minutes. This indicates that after 30 minutes of reaction time ferrous iron generation rate slightly exceeded the consumption rate. After 180 minutes, the concentration of ferrous iron became relatively steady at about 2 mg/L. Hydrogen peroxide showed different behavior than ferrous iron during the reaction. The concentration of hydrogen peroxide decreased rapidly in the first 60 minutes, and then went down slightly reaching the residual concentration of 21 mg/L in 300 minute.

3.7 Effect of Initial TOC Concentration

Because every compound in the wastewater has different removal efficiency, the overall removal efficiency can be affected by different initial concentration of the compounds. The effect of the initial concentration on the wastewater treatment efficiency was studied by diluting the original wastewater to different TOC values. Figure 7 shows the treatment results at different initial TOC values. It was very interesting to note that the observed results were opposite to what one would expect, e.g. better TOC removal at higher concentrations. TOC removal efficiency increased and reached the maximum value of 55% when the dilution factor increased to 10. Further increase in the dilution factor above 10 resulted in decrease in the TOC removal efficiency. When the initial wastewater was diluted 50 times the removal efficiency decreased to 23% and COD removal followed similar pattern. In order to explain such behavior, the average carbon oxidation number of the residual solutions was determined. When the samples were diluted 10 times, the treatment resulted in decrease of the oxidation state from -1.1 to -3.1, but at dilution factor of 50, carbon oxidation number slightly increased from -3.1 to -2.0 compared to the sample with the dilution factor of 10. At the same time, gradual accumulation of precipitate was observed at the bottom of the reactor at dilution factors greater than 10. Moreover, the quantity of precipitate increased proportionally to the dilution factor. The highest quantity of precipitate of 67 mg/L was reached after treatment by US- Fenton at the dilution factor of 50. The precipitate was collected, digested, and analyzed for TOC and iron concentration. The results showed that the precipitate contained TOC and iron in the amounts of 19.1 mg-C/L and 27.9 mg/L, respectively. The changes in the average carbon oxidation states indicated that possibly some new organic intermediates formed during the treatment process that caused TOC increase in the solution. With the initial TOC of sample decreasing, some intermediates formed slightly soluble ferrous iron complexes that easily precipitated and lead to the decrease of total iron concentration of solution and the efficiency of Fenton process, therefore TOC and COD removal became worse.

3.8 Toxicity

The toxicity of the wastewater and treated wastewater were checked by *E. coli*. The latter is easy to culture in laboratory conditions and shows a high production rate of CO_2. The final CO_2 concentration in the culture medium was measured by flow injection analysis with a conductmetric detector. Fig. 9 shows the *E. coli* responses to different sample. The results showed treatment by US- Fenton can reduce the toxicity as well as dilution. Due to hydrogen peroxide can also inhibit the respiration of *E. coli* and the concentration of hydrogen peroxide which was 1374 mg/L was still high in the treated wastewater after reaction for 60 minutes, the toxicity in the sample treated for 60 minutes was stronger than that diluted 10 times. In sample treated for 300 minutes, the inhibition in *E. coli* showed weakest. It is likely that some intermediates generated during the reaction were beneficial to *E. coli* growth.

4. Conclusion

Several methods, such as photolysis (UV), Fenton, photo-Fenton, and ultrasound-Fenton (US-Fenton), were studied for the treatment of ammunition wastewater. The results showed that US-Fenton process can be a promising method for ammunition wastewater treatment. The effect of operational parameters (ultrasonic intensity, pH, temperature, hydrogen peroxide to ferrous iron molar ratio, initial TOC concentration, and reaction time) on the TOC, COD, and color removal efficiencies were studied. Generally, the degradation of TOC, COD, and color increased with decrease in pH and increase in temperature and US intensity. During the treatment process, the optimal molar ratio of hydrogen peroxide to ferrous iron was 500. The effect of initial TOC concentration was complicated compared to other factors. The results showed that too low and high initial TOC concentration caused decline in the removal efficiency and the optimum initial TOC concentration of 70 mg-C/L.

Toxicity of the original wastewater, diluted wastewater, and treated wastewater has been investigated. The results showed that both dilution and treatment of ammunition wastewater by US-Fenton reduced the toxicity of the wastewater. The toxicity was affected by wastewater dilution factor and US-Fenton treatment duration. For shorter treatment times (i,e, 60 min), the toxicity of the wastewater treated by US-Fenton process increased compared to original wastewater diluted 10 times. This was due to the presence of residual unreacted hydrogen peroxide inhibiting *E. coli* respiration. However, after 300 minutes treatment period, the toxicity of wastewater almost disappeared.

References

[1] K. Ayoub, E.D. van Hullebusch, M. Cassir, A. Bermond, Application of advanced oxidation processes for TNT removal: A review, J. Hazard. Mater., 178 (2010) 10-28.

[2] M.J. Liou, M.C. Lu, J.N. Chen, Oxidation of explosives by Fenton and photo-Fenton processes, Water Research, 37 (2003) 3172-3179.

[3] J. Hawari, A. Halasz, T. Sheremata, S. Beaudet, C. Groom, L. Paquet, C. Rhofir, G. Ampleman, S. Thiboutot, Characterization of metabolites during biodegradation of hexahydro-1,3,5-trinitro-1,3,5-triazine (RDX) with municipal anaerobic sludge, Applied and Environmental Microbiology, 66 (2000) 2652-2657.

[4] P.M. Bradley, F.H. Chapelle, J.E. Landmeyer, J.G. Schumacher, MICROBIAL TRANSFORMATION OF NITROAROMATICS IN SURFACE SOILS AND AQUIFER MATERIALS, Applied and Environmental Microbiology, 60 (1994) 2170-2175.

[5] P.R. Gogate, S. Mujumdar, A.B. Pandit, Sonochemical reactors for waste water treatment: comparison using formic acid degradation as a model reaction, Advances in Environmental

Research, 7 (2003) 283-299.
[6] W.S. Chen, G.C. Huang, Sonochemical decomposition of dinitrotoluenes and trinitrotoluene in wastewater, J. Hazard. Mater., 169 (2009) 868-874.
[7] M.R. Hoffmann, I. Hua, R. Hochemer, Application of ultrasonic irradiation for the degradation of chemical contaminants in water, Ultrasonics Sonochemistry, 3 (1996) S163-S172.
[8] H. Yin, T.K. Wood, B.F. Smets, Reductive transformation of TNT by Escherichia coli: pathway description, Applied Microbiology and Biotechnology, 67 (2005) 397-404.
[9] K.H. Kang, D.M. Lim, H. Shin, Oxidative-coupling reaction of TNT reduction products by manganese oxide, Water Research, 40 (2006) 903-910.
[10] S.Y. Oh, P.C. Chiu, B.J. Kim, D.K. Cha, Enhancing Fenton oxidation of TNT and RDX through pretreatment with zero-valent iron, Water Research, 37 (2003) 4275-4283.
[11] G. Yardin, S. Chiron, Photo-Fenton treatment of TNT contaminated soil extract solutions obtained by soil flushing with cyclodextrin, Chemosphere, 62 (2006) 1395-1402.
[12] V.S. Moholkar, Mechanistic optimization of a dual frequency sonochemical reactor, Chemical Engineering Science, 64 (2009) 5255-5267.
[13] P. Riesz, T. Kondo, C.M. Krishna, SONOCHEMISTRY OF VOLATILE AND NONVOLATILE SOLUTES IN AQUEOUS-SOLUTIONS - EPR AND SPIN TRAPPING STUDIES, Ultrasonics, 28 (1990) 295-303.
[14] G.A.D. Briggs, ULTRASOUND - ITS CHEMICAL, PHYSICAL AND BIOLOGICAL EFFECTS - SUSLIK,KS, Interdisciplinary Science Reviews, 15 (1990) 190-191.
[15] R.A. Torres, F. Abdelmalek, E. Combet, C. Petrier, C. Pulgarin, A comparative study of ultrasonic cavitation and Fenton's reagent for bisphenol A degradation in deionised and natural waters, J. Hazard. Mater., 146 (2007) 546-551.
[16] Y.G. Adewuyi, Sonochemistry: Environmental science and engineering applications, Industrial & Engineering Chemistry Research, 40 (2001) 4681-4715.
[17] R. Kidak, N.H. Ince, Ultrasonic destruction of phenol and substituted phenols: A review of current research, Ultrasonics Sonochemistry, 13 (2006) 195-199.
[18] P.S. Bapat, P.R. Gogate, A.B. Pandit, Theoretical analysis of sonochemical degradation of phenol and its chloro-derivatives, Ultrasonics Sonochemistry, 15 (2008) 564-570.
[19] I. Gultekin, N.H. Ince, Ultrasonic destruction of bisphenol-A: The operating parameters, Ultrasonics Sonochemistry, 15 (2008) 524-529.
[20] O. Hamdaoui, E. Naffrechoux, Sonochemical and photosonochemical degradation of 4-chlorophenol in aqueous media, Ultrasonics Sonochemistry, 15 (2008) 981-987.
[21] Z.B. Guo, R. Feng, J.H. Li, Z. Zheng, Y.F. Zheng, Degradation of 2,4-dinitrophenol by combining sonolysis and different additives, J. Hazard. Mater., 158 (2008) 164-169.
[22] Z.L. Wu, B. Ondruschka, G. Cravotto, Degradation of Phenol under Combined Irradiation of Microwaves and Ultrasound, Environmental Science & Technology, 42 (2008) 8083-8087.
[23] Y. Segura, R. Molina, F. Martinez, J.A. Melero, Integrated heterogeneous sono-photo Fenton processes for the degradation of phenolic aqueous solutions, Ultrasonics Sonochemistry, 16 (2009) 417-424.
[24] I.K. Kim, C.P. Huang, P.C. Chiu, Sonochemical decomposition of dibenzothiophene in aqueous solution, Water Research, 35 (2001) 4370-4378.
[25] F. Chemat, P.G.M. Teunissen, S. Chemat, P.V. Bartels, Sono-oxidation treatment of humic substances in drinking water, Ultrasonics Sonochemistry, 8 (2001) 247-250.
[26] J.G. Lin, C.N. Chang, J.R. Wu, Y.S. Ma, Enhancement of decomposition of 2-chlorophenol with ultrasound/H2O2 process, Water Science and Technology, 34 (1996) 41-

48.
[27] R.B. Doppalapudi, G.A. Sorial, S.W. Maloney, Electrochemical reduction of simulated munitions wastewater in a bench-scale batch reactor, Environmental Engineering Science, 19 (2002) 115-130.
[28] D.K. Palaniswamy, G.A. Sorial, S.W. Maloney, Electrochemical reduction of 2,4,6-trinitrotoluene, Environmental Engineering Science, 21 (2004) 203-218.
[29] I. Grcic, M. Obradovic, D. Vujevic, N. Koprivanac, Sono-Fenton oxidation of formic acid/formate ions in an aqueous solution: From an experimental design to the mechanistic modeling, Chemical Engineering Journal, 164 (2010) 196-207.
[30] E.V. Rokhina, M. Lahtinen, M.C.M. Nolte, J. Virkutyte, The influence of ultrasound on the RuI3-catalyzed oxidation of phenol: Catalyst study and experimental design, Applied Catalysis B-Environmental, 87 (2009) 162-170.
[31] Y.W. Kang, M.J. Cho, K.Y. Hwang, Correction of hydrogen peroxide interference on standard chemical oxygen demand test, Water Research, 33 (1999) 1247-1251.
[32] W.F. Jardim, C. Pasquini, J.R. Guimaraes, L.C. Defaria, SHORT-TERM TOXICITY TEST USING ESCHERICHIA-COLI - MONITORING CO2 PRODUCTION BY FLOW-INJECTION ANALYSIS, Water Research, 24 (1990) 351-354.
[33] H.H. Sun, S.P. Sun, J.Y. Sun, R.X. Sun, L.P. Qiao, H.Q. Guo, M.H. Fan, Degradation of azo dye Acid black 1 using low concentration iron of Fenton process facilitated by ultrasonic irradiation, Ultrasonics Sonochemistry, 14 (2007) 761-766.
[34] H. Zhang, H.J. Choi, C.P. Huang, Effects of important reaction conditions on landfill leachate treatment by Fenton process, Fresenius Environmental Bulletin, 15 (2006) 43-47.
[35] I.K. Kim, C.P. Huang, Sonochemical degradation of polycyclic aromatic sulfur hydrocarbons (PASHs) in aqueous solutions exemplified by benzothiophene, Journal of the Chinese Institute of Engineers, 28 (2005) 1107-1118.
[36] J. Liang, S. Komarov, N. Hayashi, E. Kasai, Recent trends in the decomposition of chlorinated aromatic hydrocarbons by ultrasound irradiation and Fenton's reagent, Journal of Material Cycles and Waste Management, 9 (2007) 47-55.
[37] E. Psillakis, G. Goula, N. Kalogerakis, D. Mantzavinos, Degradation of polycyclic aromatic hydrocarbons in aqueous solutions by ultrasonic irradiation, J. Hazard. Mater., 108 (2004) 95-102.

Table 1. Physical chemical characteristics of ammunition wastewater

pH	Conductivity (μs/cm)	TOC	COD	TS	SS	TDS	Absorbance (348 nm)
		mg/L					
2.8±0.2	218.82	710±50	1450±150	2.24	0.01	2.16	3.873±0.01
Ionic composition, mg/L							
Cl^-	NO_3^-	SO_4^{2-}	Na^+	K^+	Ca^{2+}	Mg^2	NH_4^+
35.8	2503	46.7	157.5	13.6	29.9	0.02	219.9

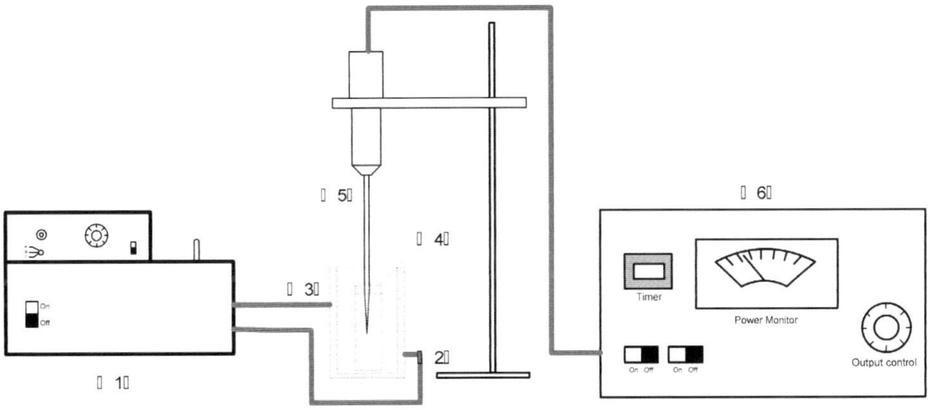

Figure 1. Schematic diagram of experimental setup for the sonochemical degradation of ammunition wastewater: (1) water circulation bath and temperature control system, (2) cooling water inlet, (3) cooling water outlet, (4) reactor, (5) converter and (6) ultrasonic generator

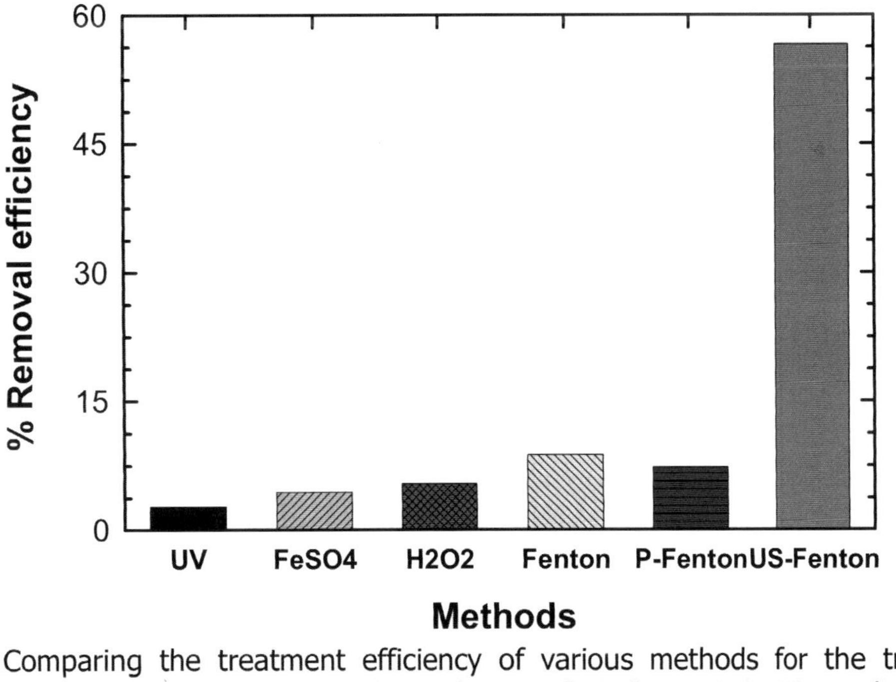

Figure 2. Comparing the treatment efficiency of various methods for the treatment of ammunition wastewater. Experimental conditions: $[TOC]_0$ = 710±50 mg/L, $[COD]_0$ = 1450±150 mg/L, Initial pH = 2.7±0.1, $[Fe^{2+}]$ = 5×10^{-4} M, $[H_2O_2]$ = 0.25 M, [NaCl] = 10^{-2} M , $UV_{254\ nm}$ irradiation = 14 watts, US intensity = 300 watts , Temperature = 25℃, Reaction time = 60 minutes.

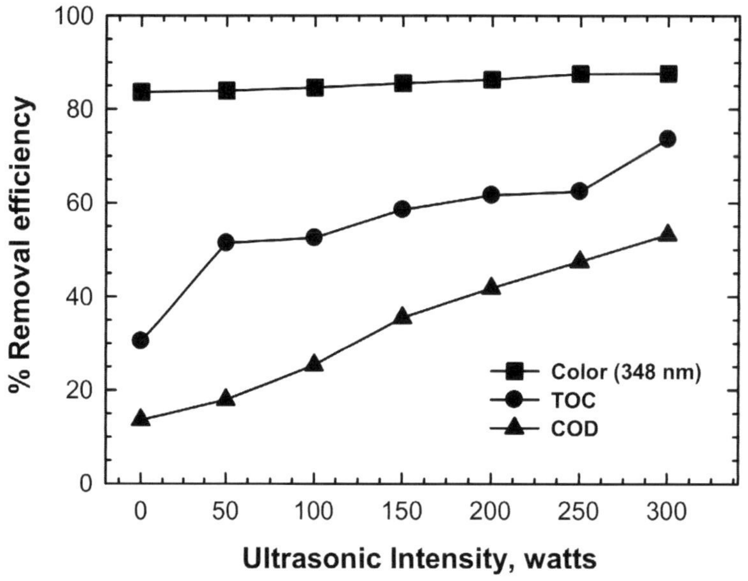

Figure 3. Effect of ultrasound intensity on the removal of TOC, COD, and color from the ammunition wastewater. Experimental conditions: $[TOC]_0$ = 710±50 mg/L, $[COD]_0$ = 1450±150 mg/L, Initial pH = 2.7±0.1, $[Fe^{2+}] = 5\times10^{-4}$ M, $[H_2O_2]$ = 0.25 M, [NaCl] = 10^{-2} M , Temperature = 25°C, Reaction time =60 minutes.

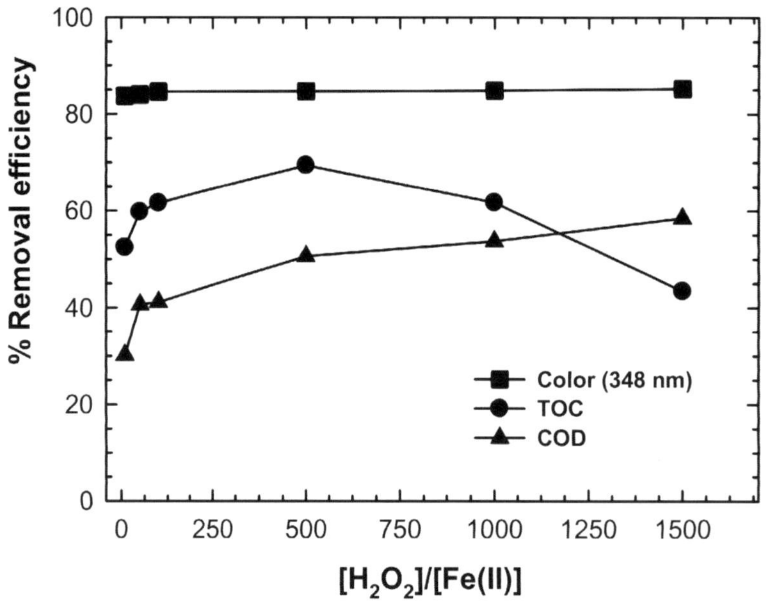

Figure 4. Effect of $[H_2O_2]/[Fe^{2+}]$ on the removal of TOC, COD, and color from the ammunition wastewater. Experimental conditions: $[TOC]_0$ = 710±50 mg/L, $[COD]_0$ = 1450±150 mg/L, Initial pH = 2.7±0.1, [NaCl] = 10^{-2} M, Temperature = 25°C, US intensity = 300 watts, Reaction time = 60 minutes.

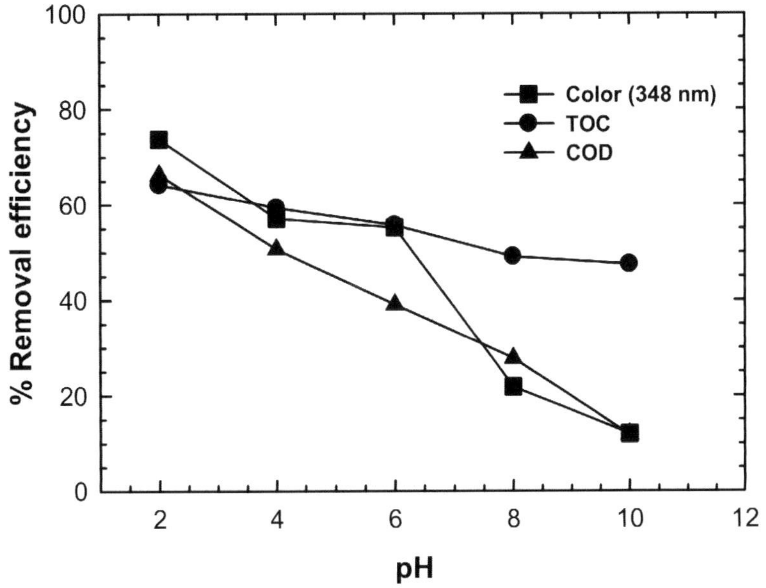

Figure 5. Effect of pH on the removal of TOC, COD, and color from the ammunition wastewater. Experimental conditions: $[TOC]_0$ = 710±50 mg/L, $[COD]_0$ = 1450±150 mg/L, $[Fe^{2+}] = 5\times10^{-4}$M, $[H_2O_2]$ = 0.25 M, [NaCl] = 10^{-2} M, Temperature = 25°C, US intensity = 300 watts, Reaction time = 60 minutes.

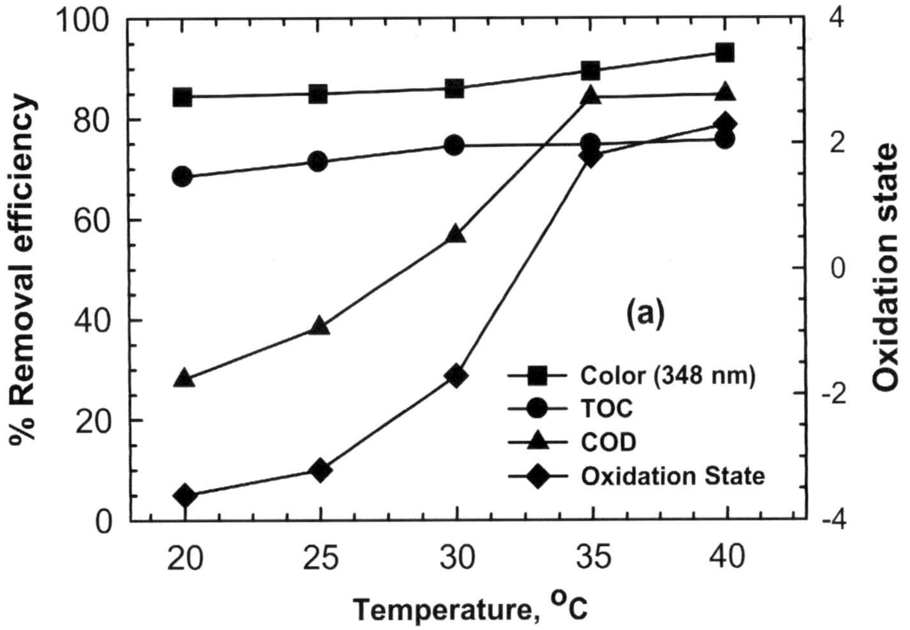

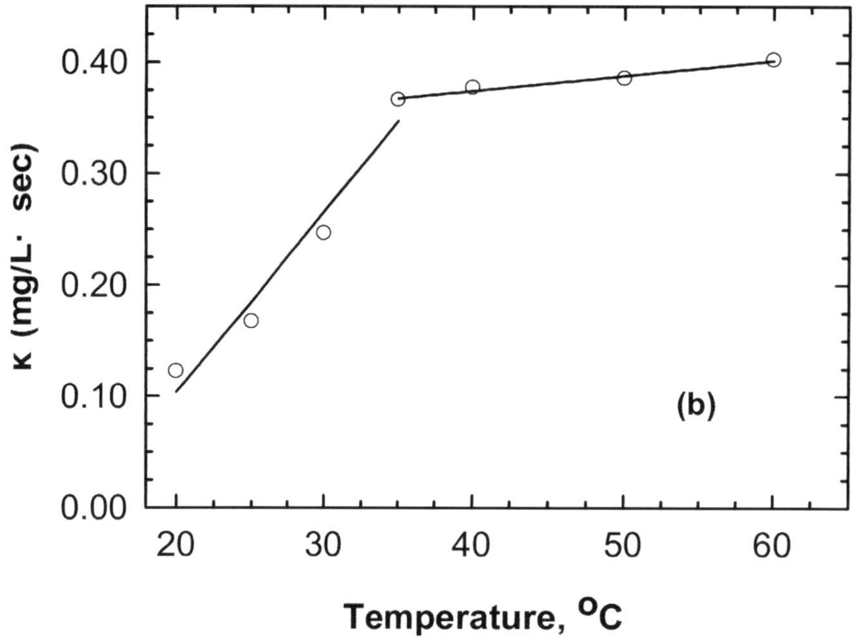

Figure 6. Effect of temperature on changes of carbon oxidation state and TOC, COD, and color removal efficiencies (a) and COD degradation rate constant for the ammunition wastewater (b). Experiment conditions: $[TOC]_0$ = 710±50 mg/L, $[COD]_0$ = 1450±150 mg/L, $[Fe^{2+}]$ = 5×10-4 M, $[H_2O_2]$ = 0.25M, [NaCl] = 10^{-2} M, Initial pH = 2.7±0.1, US intensity = 300 watts, Reaction time = 60 minutes.

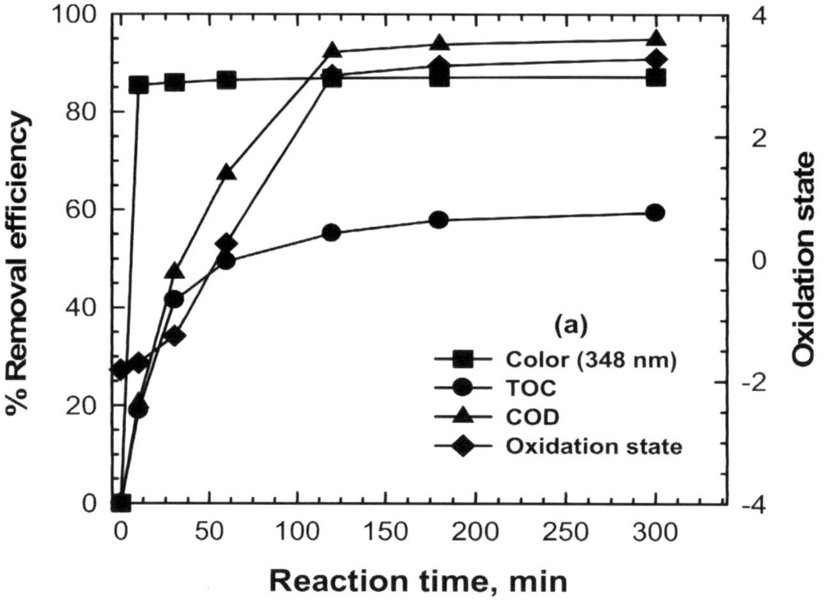

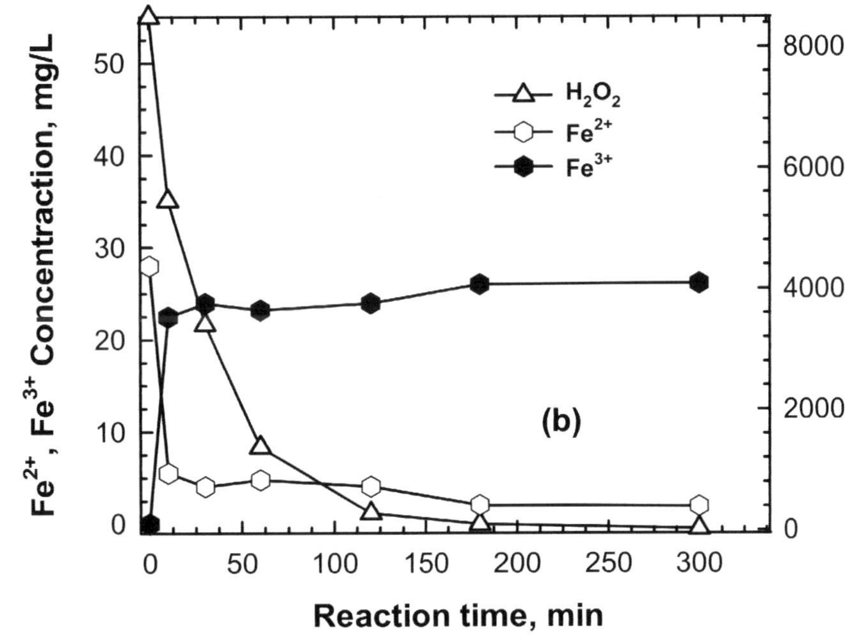

Figure 7. The removal of TOC, COD, and color from the ammunition wastewater (a) and change in the concentration of Fe^{2+}, Fe^{3+}, and H_2O_2 as a function of time (b). Experimental conditions: $[TOC]_0$ = 710±50 mg/L, $[COD]_0$ = 1450±150 mg/L, Initial pH = 2.7±0.1, $[Fe^{2+}]$ = 5×10^{-4} M, $[H_2O_2]$ = 0.25M, [NaCl] = 10^{-2} M, US intensity = 300 watts.

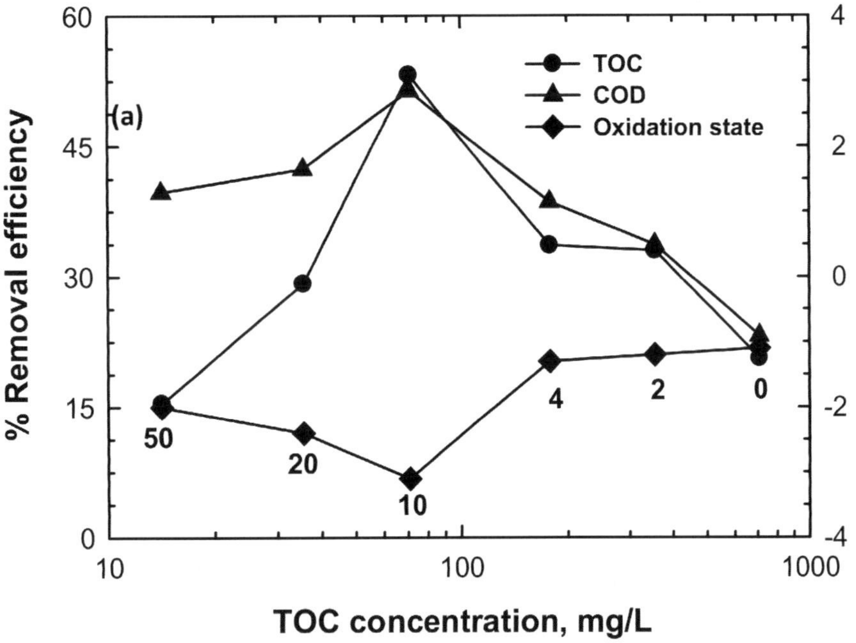

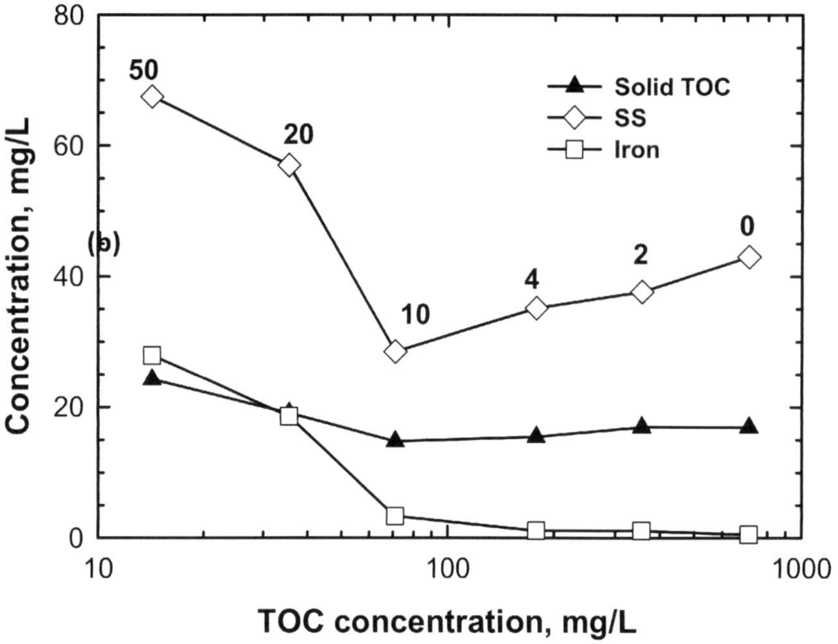

Figure 8. The effect of the initial concentration on the TOC and COD removal efficiency and carbon oxidation state during the treatment of the ammunition wastewater (a) and on the suspended solids (SS) concentration, iron concentration, and solid TOC (b). Numbers represent the dilution factor of the ammunition wastewater. Experimental conditions: $[TOC]_0$ = 710±50 mg/L, $[COD]_0$ = 1450±150 mg/L, $[Fe^{2+}]$ = 5×10^{-4} M, $[H_2O_2]$ = 0.25M, [NaCl]= 10^{-2} M, Initial pH = 2.7±0.1, Temperature = 25°C, US intensity = 300 watts, Reaction time = 60 min.

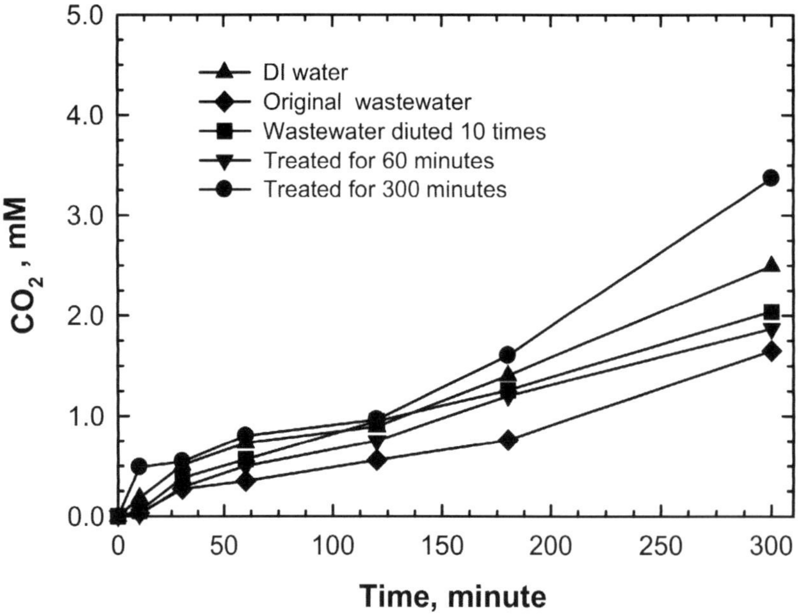

Figure 9. The toxicity test for the treated and untreated ammunition wastewater. Experimental conditions: pH = 7.0, Temperature = 37°C, Culture medium volume =50 mL.

(399b) COD Reduction From Leachate with An Advanced Oxidation Process (AOP)

Kashinath Banerjee, Veolia Water Solution & Tech, Moon Township, PA and Tapas Das, School of Engineering, Saint Martin's University, Lacey, WA

A landfill in the eastern part of United States was in operation between early to mid 1990s. The existing leachate treatment facility consists of pH adjustment, chemical coagulation, solid/liquid separation, sand filtration, carbon adsorption, and sludge dewatering. After the treatment, the water is discharged to the river, and is regulated by a National Pollution Discharge Elimination System (NPDES) permit. Occasionally, the discharge from the treatment facility exceeded the permit limitations for Chemical Oxygen Demand (COD), Biochemical Oxygen Demand (BOD), and Total Organic Carbon (TOC). The objectives of this study were to determine treatment methods which would enable compliance with the applicable discharge limitations; to establish the desired operating conditions of the process; and to investigate the effect of various parameters such as pH, catalyst dosages, reaction time, etc. on the COD destruction efficiency.

A series of influent and effluent samples were collected from the existing leachate treatment facility to characterize the wastewater, to determine the efficiency of the existing treatment system, and to perform treatability studies. All samples were preserved and handled in accordance with the U.S. EPA approved procedures. All analyses were performed in accordance with the procedures described in Standard Methods for the Examination of Water and Wastewater. Treatability studies were conducted using the following oxidation technologies: chlorine, potassium permanganate at different pH conditions, Fenton Reagent (hydrogen peroxide in presence of Ferrous ion), and Hydrogen peroxide/UV light.

The initial wastewater characterization results reveal that the COD, BOD, and TOC concentrations of the sample were approximately 150, 33, and 50 mg/L respectively. The sample had a pH of about 10.8, contained phenol ranging between 3 and 5 mg/L.

Chlorine, and potassium permanganate under acidic as well as under alkaline conditions were not effective in reducing COD to less than the discharge limitation of 35 mg/L. Treatability study results revealed Fenton Reagent or a combination of H2O2/UV which generated hydroxyl radicals (OH*) were capable to reduce the COD to below 35 mg/L within 30-minutes of reaction. The Fenton reaction was conducted at a pH of about 3.5 in presence of a hydrogen peroxide dosage of 1700 mg/L and about 200 mg/L of ferrous iron. After the reaction was completed, the residual hydrogen peroxide was removed from the reactor using a reducing agent, and the pH of the resultant mixture was raised to about 8.0 by the addition of sodium hydroxide. COD reduction with H2O2/UV is similar to that described above, except that in this process, the organic molecules will be oxidized using a combination of UV/H2O2/ catalyst.

(399c) Photocatalytic Decomposition of Geosmin and 2-Methylisoborneol by M-InVO4/TiO2 with Visible Light In Recirculating Aquaculture Systems

Sandra L. Pettit[1], Louis J. Rufo[2] and John T. Wolan[2], (1)Dept of Chemical & Biomedical Engineering, University of South Florida, Tampa, FL, (2)University of South Florida, Tampa, FL

Geosmin and 2-methylisoborneol (MIB) are odor and taste contaminants often found in drinking water and aquaculture systems. They are tertiary alcohols produced by several classes of cyanobacteria and actinobacteria. They have a strong odor which can be detected by humans at concentrations as low as 4-10 ng/L. In addition, these compounds impart a gamey, off-flavor taste to aquaculture species. Although they are not a direct threat to public health, consumers frequently use smell as a primary indicator of safety.

Conventional water treatment (coagulation/ flocculation/ sedimentation/ filtration) and chlorination are not effective in the removal of these compounds. Powdered or granulated activated carbon is generally effective. However, this usually requires high dosing rates and lose efficacy with increased water turbidity. Ozonation, although effective in drinking water systems, has several limitations in recirculating aquaculture systems (RAS). Dissolved organic material present in RAS rapidly depletes ozone, reducing the residual ozone concentration. Furthermore, ozone residuals may cause gill adhesions and increased mortality rate in fish. Because of these drawbacks, many aquaculture systems use fresh water purge to reduce geosmin and MIB concentrations. Purge systems consume large quantities of fresh water and produce significant waste streams.

Ideally, an effective removal system would balance water consumption, fish mortality, and operating cost. TiO_2 is an effective photocatalyst for the decomposition of organic compounds. It is inexpensive, photostable, and chemically and biologically inert. However, only the ultraviolet portion of the solar irradiation (amounting to ~4% of the incoming solar energy on the earth's surface) is absorbed by TiO_2 due to its high intrinsic band gap (3.2 eV for anatase and 3.0 eV for rutile). When combined with M-$InVO_4$, where M is a transition metal, photoinitiation is moved to the visible spectrum range via electron transfer. In the present study, efficacy of M-$InVO_4$/TiO_2 is explored for visible-range photocatalytic decomposition of geosmin and MIB in recirculating aquaculture systems.

See more of this Session: Photo-Catalytic and Advanced Oxidation/Reduction Processes for Water Treatment

(399d) Advanced Oxidation for Water Processing In Microstructured Reactors

Kevin Caple, Nick Kusanto, Kenneth Williamson, **Jeffrey A. Nason**, Goran N. Jovanovic and Alexandre F.T. Yokochi, School of Chemical, Biological and Environmental Engineering, Oregon State University, Corvallis, OR

An important issue in potable water processing is the need for removal of trace organic chemicals. Conventional water treatment (coagulation/ flocculation/ sedimentation/ filtration) and chlorination is not always effective in the removal of these compounds. An approach that has been proposed for the purpose is advanced oxidation employing photochemical reactions at the surface of a photocatalyst such as TiO_2 [[i]].

Since the photochemical oxidation reaction is a surface reaction, to achieve sufficient flow rates it is imperative to increase the surface to volume ratio of the reactor, and we have investigated employing microstructured reactors implementing these advanced oxidation processes. An approach that has been employed to activate reactions in macroscale reactors is the use of electric discharge through the reaction medium which leads to the creation of highly reactive species as a consequence of the electrical discharge [[ii]] as an alternative to photocatalyzed approaches. We have therefore developed a microreactor employing a low voltage corona discharge for the disinfection and organic compound oxidation in water, a technique that moves the oxidation reaction from the surface of the reactor to the bulk of the fluid, thus increasing process efficiency.

In our presentation we will discuss recent results of the use of microstructured reactors for the advanced oxidation of organic contaminants in water.

[i] Ollis, D.F.; Pelizzetti, E.; Serpone , N. "Destruction of water contaminants" *Environ. Sci. Technol.* **1991**, *25*, 1522-1529.

[ii] Kirkpatrick, M. J.; Locke, B. R. "Hydrogen, Oxygen, and Hydrogen Peroxide Formation in Aqueous Phase Pulsed Corona Electrical Discharge" *Ind. Eng. Chem. Res.* **2005**, 44, 4243-4248.

See more of this Session: Photo-Catalytic and Advanced Oxidation/Reduction Processes for Water Treatment
See more of this Group/Topical: Topical 1: Water Technology for Developed and Developing Countries (see also Separations Division)

KINETICS OF OXIDATION OF DYES IN EFFLUENTS USING A POROUS ELECTRODE OZONATOR

Alexander P. Mathews[1] and Kishora K. Panda[2]
[1]Department of Civil Engineering, Kansas State University, Manhattan, KS 66506, USA
[2]HDR Engineering, Inc., 3230 El Camino Real, Suite 200, Irvine, CA 92604

ABSTRACT

Ozonation of dye Reactive Blue 19 wastewater was conducted using an *in situ* porous electrode ozone generator and reactor system. In this system, corona discharge is made at the gas liquid interface inside a porous barrier that separates the generator from the tubular reactor. The generated ozone from the outer generator diffuses through this porous barrier and reacts with the contaminants in the fluid that is being treated inside the reactor. Oxidation studies indicate that feed gas flow rate, and ozone concentration, pH, and initial dye concentration affects the kinetics of dye decolorization. Fast kinetic regime observed for dye oxidation followed by intermediate regime for the reaction of ozone with the byproducts. The specific power consumption for ozonation of wastewater in *in situ* reactor is significantly lower than that for conventional ozonators. High mass transfer and low power consumption for this reactor system makes it an efficient and cost-effective reactor for ozonation process.

INTRODUCTION

The treatment and disposal of dye wastewater is an environmental concern due to the inadequacy of the treatment methods in treating such complex wastewater, the aesthetic problem in the receiving water, and the effect on public health. Treatment alternatives available for dye wastewater include coagulation, chemical oxidation, activated carbon adsorption, and biological treatment. Coagulation and carbon adsorption may remove certain categories of dye up to 90% but possesses the drawback of generating large amount of sludge. Conventional biological processes are not successful since most dyes are resistant to biological degradation and toxic to the organisms used (Gardner et al., 1990; Zhou and Smith, 1997). Studies have reported that partial oxidation followed by biodegradation is a cost–effective treatment alternative for this kind of wastewater (Alvares et al., 2001; Zhou and Smith, 1997).

Ozone, an oxidant of choice for many reasons in comparison to others, has limited use at present for the reason of treatment cost. Ozonation efficiency is affected due to mass transfer limitations, and ozone decomposition and reaction kinetics in the reactors. The high costs associated with ozone application can be reduced by increasing the mass transfer rate in the dissolution of ozone gas, and by reacting ozone as it is produced with the contaminants in the solution by developing an *in situ* ozone generator (Mathews et al., 2004).

Ozone generation through the electrical discharge process produces, besides ozone, highly reactive short-lived species near the corona discharge region (Sano et al., 2002). The gaseous species that are generated include free electrons, negative ions, positive ions, and uncharged radicals (Tamon et al., 1995). The current ozonation systems are unable to utilize these short-lived species for ozonation process since the reactor is separated from the generator. Sharma et al., 2000 and Sano et al., 2002 studied the impact of these species on the degradation of organic matters. They reported that the rate of decomposition of organics was higher in comparison to ozone oxidation alone, and attributed this to the reaction of organics with the short-lived species formed near the corona discharge region.

In this research, a porous electrode ozonator has been designed in which the ozone generator and the reactor form a single unit (Mathews et al., 2004). The design consists of an outer peripheral ozone generation region and an inner tubular reactor separated by a porous tube. The corona discharge occurs when a pulsed-modulated AC HT (5kV – 10kV) is applied between the outer electrode and the inner porous electrode. Generated ozone from the corona region diffuses through the porous tube into the reactor through which treatment fluid passes. This design was envisaged to provide several benefits such as increasing ozone mass transfer rate by using a fine porous diffuser, employing the reactive species generated near corona region for degradation of organic matter, consuming less power for operation, and minimizing the loss of ozone by diffusing immediately to water upon generation. The specific goal of the present work was to study the decolorization kinetics of Reactive Blue 19 dye in the tubular *in situ* ozonator, and to determine the effects of process parameters on ozonation efficiency and power consumption in comparison with conventional ozonators.

METHODOLOGY

In situ ozone generator and reactor design

The porous electrode *in situ* ozone generator and reactor system was designed essentially in one unit wherein the inner tubular reactor is separated from the outer peripheral concentric ozone generator by a porous barrier. An outer glass tube of length 29.6 cm and internal diameter 2.62 cm was placed concentrically with an equal length inner porous ceramic tube of outer diameter 2.52 cm, and internal diameter 1.95 cm, and the peripheral gap was made airtight by sealing the ends. The volume of the inner tubular reactor was 88.4 cm_3. Details of the reactor system is described elsewhere (Mathews et al., 2004). Ozone is generated by electrical corona discharge process. The electrical circuitry was designed to supply less hazardous high frequency (15 to 20 kHz) pulsed modulated HT AC (5KV to 10 KV) from a regular 110 V wall terminal (Padmanabhan et al., 2000). The outer glass tube covered with aluminum foil was used as the hot electrode, and the porous tube with an embedded steel wire acted as the ground electrode. Corona discharge occurred in the peripheral gap when high voltage is applied between the two electrodes.

Experimental

Figure 1 shows the experimental set up for studying the oxidation of dye wastewater in the *in situ* ozone generator and reactor system. The dye oxidation process was studied under a closed loop water recirculation system wherein the water was continuously pumped from a recirculation unit of 1.1L capacity and circulated through the reactor. Oxygen (ultra high purity and extra dry grade, Air Products, KS, U.S.A) was supplied through a flow meter with needle valve to the generator for ozone generation. The ozone and oxygen mixture from the generator diffused through the porous ceramic tube and absorbed into the flowing wastewater inside the reactor.

The unreacted dye and the unabsorbed gas were carried to the recirculation unit where the gas was separated and escaped to the vent as off-gas. The unit was designed to separate gas in a disengagement section at the top of the unit while holding the liquid at the bottom. The contents of the recirculation unit were mixed with a magnetic stirrer. The recirculation unit had openings for taking water samples and adding make-up water, off-gas line, and temperature probe. Samples were taken from the recirculation unit and analyzed for dissolved ozone and dye concentration. The off-gas from the reactor was separated from the liquid in the recirculation unit and then sent to an online ozone gas analyzer (BMT 964 analyzer, BMT Messetechnik, Berlin, Germany) through a demister. The off-gas was absorbed in the 2% KI trap and the gas was vented out.

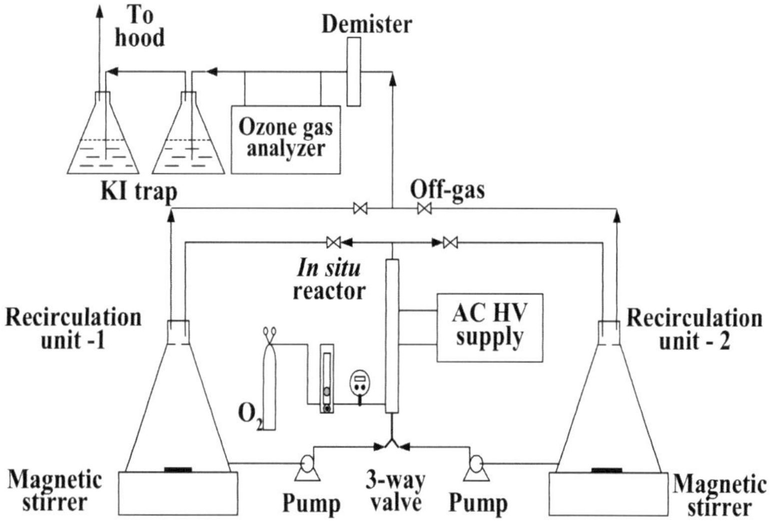

Figure 1. Experimental set-up for dye oxidation study using *in situ* reactor

Analytical Methods

Ozone concentration in the inlet gas was determined indirectly as the generated ozone diffuses and dissolves into the liquid immediately. Ozone generation rates were determined from the analyzed steady state dissolved and off-gas ozone concentrations. Eq.1 can be used to calculate the ozone generation rates.

$$G_{in} = G_{OSS} + \frac{kC_{USS}^{m}V}{Q_g} \tag{1}$$

Here G_{OSS}, C_{USS} are the steady state concentration of off-gas and dissolved ozone respectively, k the ozone self-decomposition rate constant, V the total volume of water in the recirculation unit and the reactor, Q_g the gas flow rate, and m the reaction order for the self-decomposition of ozone in water.

Prior to the dye oxidation experiment, 0.5L of distilled, deionized water at pH 2 was absorbed with ozone in the reactor by circulating the water from another recirculation unit placed in parallel (Fig.1). The experiment was carried out until dissolved ozone concentration in the recirculation unit reached a constant value. A voltage regulator (Variac Autotransformer, General Radio Company) was used to adjust the power supply to the circuitry to obtain a desired generation rate. The dissolved ozone concentration was determined by using the indigo colorimetric method (Clesceri et al., 1998).

Dye concentration in wastewater was determined from the standard curve developed for Reactive Blue 19 dye. Standard dye solutions at concentrations of 0.002 to 0.16 g/L were prepared, and their absorbance was measured at the maximum absorbed wavelength of 590nm on the spectrophotometer (Cary 50 Bio UV-Vis system, Varian, Walnut Creek, California). During dye ozonation experiments, water samples were removed from the recirculation unit and absorbance was read immediately on the spectrophotometer without any dilution. The sample water was returned into the recirulcuation unit to maintain a constant liquid volume inside the unit.

Distilled deionized (DDI) water (conductivity <0.06 μS/cm) was used to prepare dye wastewater. Oxidation of dye Reactive Blue 19 (Sigma-Aldrich Co., St.Louis, MO) was studied at pH 2 and 7. The pH was adjusted to 2 by addition of sulfuric acid, and to 7 by addition of calculated amounts of NaH_2PO_4 and Na_2HPO_4. All experiments were conducted at a temperature of 21 ± 0.5^0C, and ionic strength of 0.05 M (pH 2) and 0.1M (pH 7). Other chemicals used were reagent grade and were obtained from either Fisher Scientific or Sigma-Aldrich Co.

RESULTS AND DISCUSSIONS

Ozone generation rate

The ozone generation rate was calculated from dissolved and off-gas ozone measurements. It was assumed that the ozone generated in the corona discharge region was either absorbed into the liquid phase or escaped into the gas phase, and there is no ozone loss by reaction while diffusing through the porous ceramic electrode. Ozone generation rates at different gas flow rates and corresponding steady state off-gas ozone concentrations are shown in Table 1. As indicated in the Table and Eq. (1), the difference between G_{in} and G_{OSS} is due to the self-decomposition of ozone in water, and varies from 0.529% to 4.332%, increasing with a decrease in flow rate. The differences are higher at a higher pH of 7 due to the higher ozone decomposition rate by reaction with the hydroxyl ions. The generation rates were also verified by using the KI wet chemistry method (Rakness et al., 1996) and results were found to be within ±5% of the values (Table 1) calculated by the mass balance method. This deviation is acceptable considering the fluctuations in ozone generation using high frequency and high voltage corona discharge.

Ozonation of dye in in situ reactor

The concentration of dye, dissolved ozone, and off-gas ozone in recirculation unit was monitored during the ozonation of Reactive Blue 19 dye. Figure 2 shows the profiles for the three measured parameters for an ozonation experiment in *in situ* reactor. From the figure, it can be concluded that the Reactive Blue 19 ozonation can be divided into two phases. During the first 35 minutes, the process is governed by the oxidation of dye. The reaction regime is observed to be fast as the

Table 1 - Estimated ozone generation rate based on mass balance (pH = 2, and liquid velocity u_l= 20.1 m/h) and KI test (Panda and Mathews, 2004).

Feed gas flow Q_g (L/min)	Feed gas pressure (kPa)	Steady-state off-gas O_3 G_{OSS} (mg/L)	Ozone generated* G_{in} † (mg/L)	Steady-state off-gas O_3 flow rate (g/h)	Generation rate (g/h) †		Variation (%)
					Mass Balance	KI Test	
0.02	115.92	35.00	36.58	0.0493	0.0515	0.0542	0.84
0.03	119.37	31.25	31.88	0.0680	0.0694	0.0671	1.23
0.04	122.82	25.00	25.54	0.0747	0.0763	0.0737	1.70
0.05	133.16	33.80	33.98	0.1368	0.1375	0.1399	2.24

† not at constant voltage
* Calculated from mass balance

dissolved ozone concentration in water is zero. All the absorbed ozone in the water reacted with the dye either in the film or in the bulk water. The chemical reaction in the liquid increased the mass transfer rate which was observed from the difference between the feed gas and off-gas ozone concentration. During the second period i.e after 35 minutes, the oxidation of by-products governs the oxidation process. The reaction kinetics of by-products with ozone in this phase is low and not limited by the mass transfer rate. In addition, appearance of dissolved ozone indicates that there is no significant reaction of ozone with the by-products inside the film.

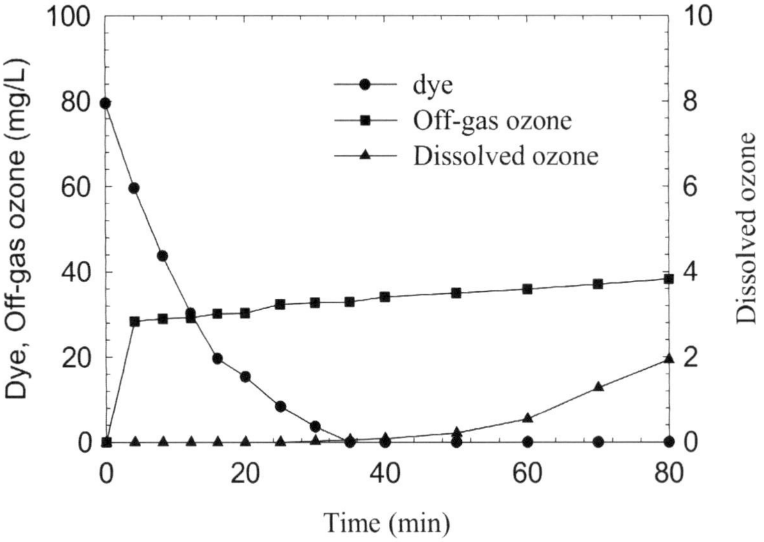

Figure 2. Concentration profiles for dye, dissolved, and off-gas ozone in *in situ* reactor (pH=2, G_{in} =50 g/Nm^3, C_o = 80 mg/L, Q_g=50 mL/min)

Effect of gas flow rate

The rate of dye oxidation was investigated by varying the gas flow rate. Figure 3 presents the dye decolorization at two different gas flow rates of 56 mL/min and 118 mL/min for an initial dye concentration of 80 mg/L. The rate of decolorization was higher at gas flow of 118mL/min than that at 56 mL/min. Since dye oxidation reaction is in fast kinetic regime until dye is present in water, the decolorization rate is mass transfer limited. So at a constant ozone generation rate, higher reaction rate is due to the increased mass transfer rate at gas flow rate of 118 mL/min in comparison to that at 56 mL/min. Gas flow rate has been reported to increase the specific interfacial area due to decrease in bubble size in stirred tank reactor (Sheffer and Esterson, 1982; Wu and Masten, 2001) and bubble columns (Roustan et al., 1996). In the *in situ* reactor, the mixing regime and the bubble size are possibly changed with increase in gas flow rate, and that affected the overall mass transfer rate by decreasing the bubble size and increasing the mixing intensity inside the reactor. The bubble diameters were difficult to measure in this small and closed reactor, but mixing was observed to increase with the increase in gas flow.

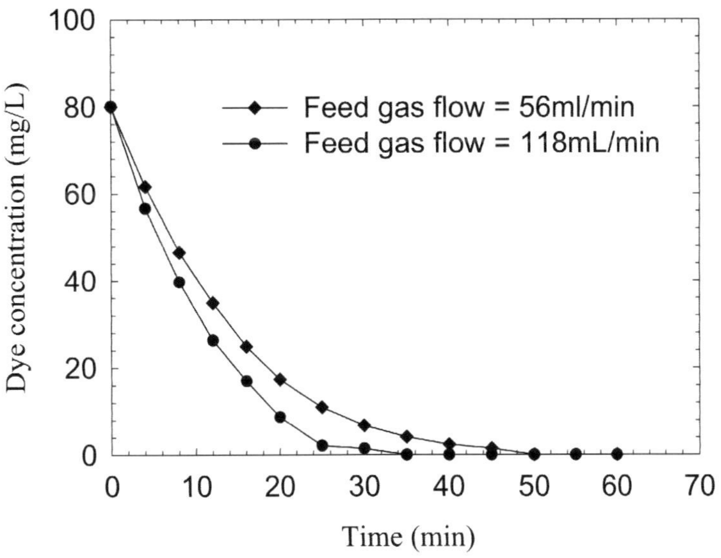

Figure 3. Effect of feed gas flow rate on dye decolorization (C_o = 80 mg/L, G_{in} = 50 g/Nm^3, and pH = 2)

From the mass transfer studies in *in situ* reactor, a power law relationship, given in Equation 2 and 3, has been observed between superficial gas velocity and overall mass transfer coefficient (Panda and Mathews, 2004).

$$\text{pH} = 2\text{: } K_La = 0.4149\, u_g^{0.3878} \quad (2)$$

$$\text{pH} = 7\text{: } K_La = 0.3775\, u_g^{0.4495} \quad (3)$$

where K_La is the overall mass transfer coefficient (min^{-1}), and u_g the superficial gas velocities (m/h).

Effect of pH of water

The ozonation of dye at initial pH of 2 and 7 was examined, and the results for an experimental condition with initial dye concentration of 40 mg/L and ozone generation rate of 60.6 g/Nm^3 are shown in Figure 4. The rate of dye decolorization was observed to be higher at pH 2 than that at pH 7. Ozonation reaction pathways include direct oxidation by ozone or radical oxidation by hydroxyl radical. The hydroxyl radical reaction predominates under basic conditions and oxidation potential of the reaction is much higher than that for molecular ozone. So reaction rate at pH 7 is expected to be higher than at pH 2. For this particular anthraquinone dye, similar findings were reported in literature in relation to the effect of pH on dye oxidation (Chu and Ma, 2000). The characteristics of reactive blue 19 such as insensible structure for oxidation, leuco-form, and higher solubility in acidic water contribute to a higher decolorization rate in acidic conditions.

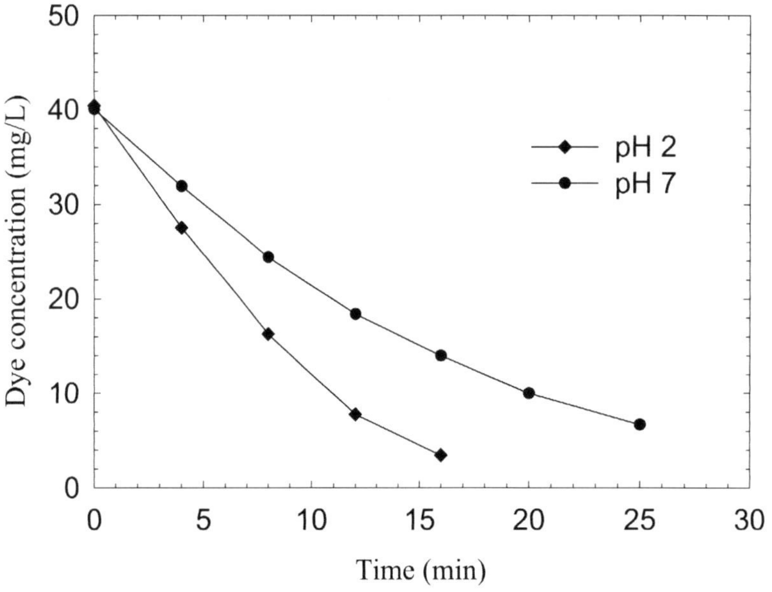

Figure 4. Effect of pH on decolorization rate of dye wastewater (C_o =40 mg/L, G_{in} = 60.6 g/Nm^3, and Q_g =33.6 mL/min)

Effect of feed gas ozone concentration

Figure 5 displays the effect of feed gas ozone concentration on the rate of decolorization at an initial dye concentration of 80 mg/L and pH 2. Three ozone concentrations investigated were

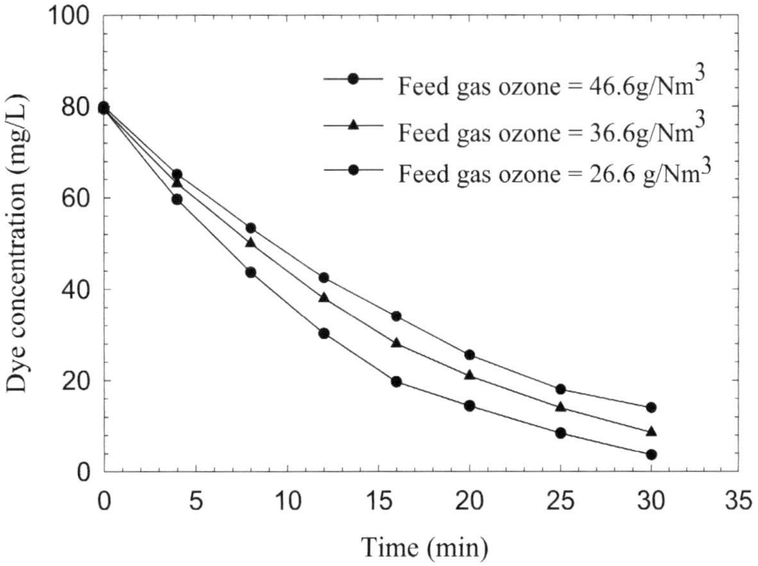

26.6 g/Nm^3, 36.6 g/Nm^3, and 46.6 g/Nm^3 at a constant feed gas flow of 50 mL/min. As shown in the figure, the decolorization rate increased with the increase in the feed gas concentration. Increasing feed gas ozone concentration increases the mass transfer rate, thus increasing the decolorization of dye for which reaction rate is mass transfer limited until complete dye decolorization. Shu and Huang, 1995 reported the increase in oxidation rate of dye Acid Orange 10 with an increase in mass transfer rate in the reactor. Decolorization of dye reactive blue 19 was studied by Lall et al., 2003, and observed an increase in reaction rate with the increase in concentration of ozone in inlet gas.

Figure 5. Effect of feed gas ozone concentration on dye decolorization (C_o = 80 mg/L, pH = 2, and Q_g = 50 ml/min)

Effect of initial dye concentration

The dye decolorization in *in situ* reactor was studied by varying the initial dye concentration in water. Results of dye decolorization at two initial concentrations of 40 mg/L and 80 mg/L are presented in Figure 6. The reaction rate decreased with the increase in the initial dye

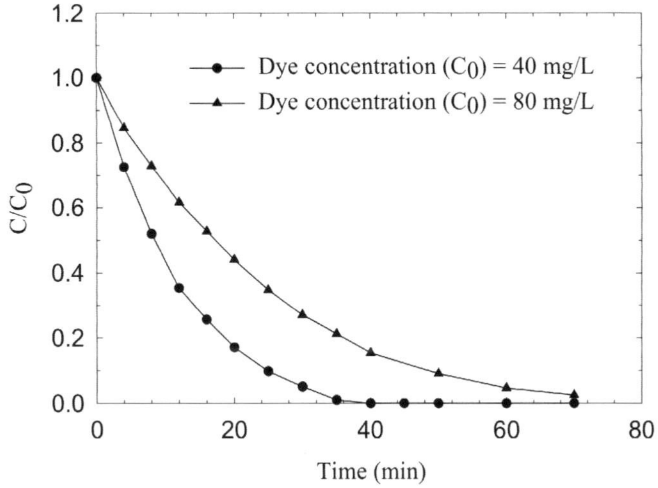

Figure 6. Effect of initial dye concentration on dye Decolorization (G_{in} = 50 g/Nm^3, pH 7 and Q_g = 50 mL/min)

concentration and this result is in agreement with the literature (Chu and Ma, 2000; Lall et al., 2003; Wu and Wang, 2001). These researchers used regular ozonators to study the dye decolorization. Hence the effect of initial dye concentration on kinetics of dye decolorization as observed in *in situ* reactor is similar to that in regular ozonators.

Power consumption

The *in situ* ozonator developed in this study is relatively small in size in comparison to regular ozonators. In comparision, *in situ* reactor needs significantly less power, required to circulate water through a small reactor. The specific power requirement is calculated for the *in situ* reactor to range from 0.002 to 0.004 kWm^{-3} (Panda and Mathews, 2004). The range of values for power requirement in bubble columns and stirred tank reactors reported in literature are from 0.01 to 0.1 kW m^{-3} and 0.5 to 10 kW m^{-3} respectively (Panda and Mathews, 2004).

CONCLUSIONS

Reaction efficiencies for dye decolorization are higher in this reactor system due to *in situ* ozone dissolution and reaction at the porous electrode gas-liquid interface, and involvement of reactive species formed near corona region in oxidation reaction. The reaction kinetics was observed to be in fast regime until dye was present in the water. Feed gas flow rate and ozone concentration, initial dye concentration, and pH of water affected the kinetics of oxidation. Kinetics of decolorization is directly dependent on the dye concentration. Higher reaction rate was observed in acidic medium than in neutral water due to the typical characteristics of this particular dye such as leuco-form, high solubility in acidic water, and insensible structure for oxidation. Small size of the reactor and absence of any stirring requirement makes this design less power intensive in comparison to the bubble column and the stirred tank reactor used in ozonation process.

ACKNOWLEDGEMENTS

This research was supported by a research grant and REU supplement from the National Science Foundation (BES – 0209343). We acknowledge and appreciate the assistance of Jamie Klein in the conduct of studies and laboratory analysis.

REFERENCES

Zhou, H.; Smith, D. W. (1997) Process Parameter Development for Ozonation of Kraft Pulp Mill Effluents. *Water Sci. Tech.*, **35** (2-3), 251.

Gardner, D. A.; Holdsworth, T. J.; Shaul, G. M.; Dostal, K. A.; Betowski, D. L. (1990) *Aerobic and Anaerobic Treatment of C.I. Disperse Blue* 791 EPA-600/S2-89/051; Office of Research and Development, U.S. Environmental Protection Agency, Cincinnati, Ohio.

Alvares, A. B. C.; Diaper, C., and Parsons, S. A. (2001) Partial Oxidation of Hydrolyzed and Unhydrolyzed Textile Azo Dyes by Ozone and the Effect on Biodegradability. *Trans IchemE,* **79** (B), 103.

Mathews, A. P.; Panda, K. K.; Ananthi, S.; Padmanabhan, K. (2004) Mass Transfer and Oxidation Kinetics in an *In Situ* Ozone Generator. *Water Sc. & Tech.*, **49** (4), 19.

Sano, N.; Kawashima, T.; Fujikawa, J.; Fujimoto, T.; Kitai, T.; Kanki, T. (2002) Decomposition of Organic Compounds in Water by Direct Contact of Gas Corona Discharge: Influence of Discharge Conditions. *Ind. Eng. Chem. Res.*, **41**, 5906.

Tamon, H.; Mizoata, H.; Sano, N.; Okazaki, M. (1995) New Concept of Gas Purification by Electron Attachment, *AIChE J.*, **41**, 1701.

Sharma, A. K.; Camaioni, D. M.; Josephson, G. B.; Goheen, S. C. (2000) Destruction of Pentachlorophenol Using Glow Discharge Plasma Process. *Env. Sci. Technol.*, **34**, 2267.

Padmanabhan, K.; Ananthi, S.; Patel, K. (2000) Portable Ozone Generator. *Electronics for You,* **32**, 46.

Clesceri, L. S.; Greenburg, A. E.; Eaton, A. D. (1998) *Standard Methods for the Examination of Water and Wastewater.* 20th Edn., Washington, DC.

Panda, K. K.; Mathews, A. P. (2004) Ozone Mass Transfer in an *In Situ* Ozone Generator and Reactor II. Experimental Studies. *Water Research,* (under review)

Sheffer, S.; Esterson, G. L. (1982) Mass Transfer and Reaction Kinetics in the Ozone/Tap Water System. *Water Research,* **16**, 383.

Wu, J. J.; Masten, S. J. (2001) Mass Transfer of Ozone in Semi Batch Stirred Reactor. *J. Env. Eng.*, **127** (12), 1089.

Roustan M.; Wang R.Y.; Wolbert, D. (1996) Modeling Hydrodynamic and Mass Transfer Parameters in a Continuous Ozone Bubble Column. *Ozone Sci. Eng.*, **18**, 99.

Chu, W.; Ma, C. (2000) Quantitative Prediction of Direct and Indirect Dye Ozonation Kinetics. *Water Res.,* **34** (12), 3153.

Shu, H. Y.; Huang, C. R. (1995) Degradation of Commercial Azo Dyes in Water Using Ozonation and UV Enhanced Ozonation Process. *Chemosphere*, **31**, 3813.

Lall, R.; Mutharaan, R.; Shah, Y. T.; Dhurajati, P. (2003) Decolorization of the Dye, reactive Blue 19, Using Ozonation, Ultrasound, and Ultrasound-Enhanced Ozonation. *Water Env. Res.*, **75**(2), 171.

Wu, J.; Wang, T. (2001) Ozonation of Aqueous Azo Dye in a Semi-Batch reactor. *Wat. Res.*, **35** (4), 1093.

(399f) Adsorptive-Reactive Nanoparticulate Systems for the In Situ Remediation of Chlorinated Hydrocarbons

Bhanukiran Sunkara[1], Jingjing Zhan[1], Yingqing Wang[1], Vijay T. John[2], Jibao He[3] and Gary McPherson[4], (1)Department of Chemical and Biomolecular Engineering, Tulane University, New Orleans, LA, (2)Chemical & biomolecular engineering, Tulane university, New orleans, LA, (3)Central Instrumental Facilities, Tulane University, New Orleans, LA, (4)Department of Chemistry, Tulane University, New Orleans, LA

Chlorinated hydrocarbons such as trichloroethylene (TCE) form a class of dense non-aqueous-phase liquid (DNAPL) toxic contaminants in soil and groundwater. The *in situ* injection of nanoscale zerovalent iron (NZVI) to reduce DNAPLs is a potentially simple, cost-effective, and environmentally benign technology that become a preferred method in the remediation of these compounds. However, unsupported NZVI particles exhibit ferromagnetism leading to particle aggregation and a loss in mobility through the subsurface. This work focuses on engineering submicron spherical carbon particles as effective carriers/supports for NZVI particles to address the *in situ* remediation of chlorinated hydrocarbons. We describe here a simple one-step process, demonstrating that it is possible to vary the placement of iron nanoparticles on carbon microspheres either on the surface of the microsphere or in the interior. These composites have high reactivities coupled with strong adsorption capabilities. The optimal size of these multifunctional particulate systems enables effective transport through sediments. The particulate systems are obtained from inexpensive precursors and through a semi-continuous method which allows for large scale synthesis of the composites necessary for eventual *in-situ* application. The detailed characterization of these multifunctional colloids through high resolution electron microscopy and their functional properties will be described. The mobility of these composites through model sediments using capillary and column transport experiments will be discussed.

See more of this Session: Photo-Catalytic and Advanced Oxidation/Reduction Processes for Water Treatment
See more of this Group/Topical: Topical 1: Water Technology for Developed and Developing Countries (see also Separations Division)

(465a) A Different Look At Water: Water for Cities and the Water-Energy Nexus

Snehal Desai, Marketing, Dow Water and Process Solutions, Edina, MN

Cities in the Developed and Developing World are Water-challenged

It should be no surprise that the demand for water outstrips its ready supply everywhere around the world. As global population soars, food, energy and freshwater are becoming increasingly scarce. Water, whether for potable or industrial use, is limited, and some supplies are not useable. Under an average economic growth scenario and without efficiency gains, global water requirements will grow from 4,500 billion cubic meters today to nearly 7,000 billion cubic meters—more than half of all the water in Lake Superior and a 50% increase in only twenty years.

By 2030, some analysts predict that available water supplies will satisfy only 60% of demand.2 According to the World Economic Forum, nearly 60% of the world's population will be living in cities by that time, causing a shortage of clean water for people and business in the urban environment worldwide.3 In that time, a third of humanity will have only half the water required to meet basic needs,4 which is likely to impact food production through its effect on agriculture which accounts for more than 70% of water usage.5 Clearly, this scenario poses serious challenges to local communities wherever they may be.

In the developing world, the challenge to municipalities is making the limited amount of fresh water clean enough to drink for millions. In no place is this playing out more intensely than China, which by 2025 will build 221 cities with a million or more people and fifteen mega cities with populations of over 10 million. Today China has approximately 300 million people with no access to water. Even though China has 6% of the world's total water resources, its large population means that the country only has 25% of the world's average water resources per capita. The UN lists China as one of 13 countries that is experiencing serious water scarcity. Of the 661 cities in China, 33% are scarce of water, while 17% of China is regarded as badly scarce of water.6 The World Health Organization states that 1/6 of the world's population does not have access to safe water for drinking. This translates into 1.1 billion people globally who do not have access to clean drinkable water.

While the challenges in the developing world seem to overshadow water issues in places like the United States and Europe they are just as real. In established cities where infrastructure already exists, municipal planners are faced with aging pipes and pumps, falling water tables and new regulations that place an additional cost burden on already stretched resources for treating current water supplies.

See more of this Session: Membrane-Based Technologies for Water and Energy Reduction

(465b) Reducing Energy Consumption In Reverse Osmosis Desalination: Cyclic or Multi-Stage Operation?

Aihua (Richard) Zhu, Chemical and Biomolecular Engineering, UCLA, Los Angeles, CA, Panagiotis D. Christofides, Department of Chemical and Biomolecular Engineering, University of California, Los Angeles, Los Angeles, CA and Yoram Cohen, Chemical & Biomolecular Engineering, University of California, Los Angeles, Los Angeles, CA

Current commercially available reverse osmosis membranes are of sufficiently high permeability to enable RO desalination process operation near the thermodynamic limit while maintaining a practical level of permeate flux. As a result, the minimum specific energy consumption (Energy consumed per unit product water) is dictated by the water recovery and not membrane permeability. However, plant footprint can be reduced since higher flux can be attained with a lower membrane surface area. As concluded in our previous work [1], significantly reduction in the cost of RO desalination is more likely to arise from optimization of process configuration (e.g. [2], [3], [7]) and control schemes [4] (e.g., to account for feed salinity fluctuation [5] and even temporal fluctuation of electrical energy purchasing price), utilization of low cost renewable energy sources, as well as more effective and lower cost feed pretreatment and brine management [6]. For example, recent developments have suggested that cyclic (semi-batch) RO operation can result in lower energy consumption relative to continuous RO operation. Also, it has been proposed that the use of multi-stage RO operation (with booster pumps between RO stages) can lead to significant energy savings by allowing RO operation that enables closer approach toward thermodynamic process reversibility. Accordingly, in the present work we examine the above two approaches for reducing RO energy consumption. During cyclic RO, the full brine stream is mixed with the fresh feed to the process and the resulting stream enters the RO module. As a result, the energy of the brine stream is utilized while the salinity of the feed in the RO module increases with time, necessitating the increase of the feed side pressure to maintain a constant water production rate. Furthermore, the RO module is taken offline periodically to remove the salt accumulated in the RO module. In order to evaluate the energy consumption for such a process, a novel computational framework was developed for computing the SEC of the RO processes with full cyclic operation accounting for the impact of downtime. In contrast, multi-stage which allow continuous permeate production enable lower energy consumption relative to a single-stage (compared at the same permeate recovery), albeit at the expense of higher membrane surface area requirement and possibly higher capital cost. The results of the present analysis demonstrate that under certain operating conditions cycling or continuous operation (with and without energy recovery devices) can be favorable from an energy viewpoint and can lead to a reduction in the cost of water production by RO desalination.

References:

1. Zhu, A., Christofides, P. D. and Cohen, Y. On RO Membrane and Energy Costs and Associated

Incentives for Future Enhancements of Membrane Permeability, JMS, 2009, 344(1-2): 1--5.
2. Zhu, A., Christofides, P. D. and Cohen, Y. Effect of Thermodynamic Restriction on Energy Cost Optimization of RO Membrane Water Desalination, IECR, 2009, 48(13): 6010-6021.
3. Zhu, A., Christofides, P. D. and Cohen, Y. Minimization of Energy Consumption for a Two-Pass Membrane Desalination: Effect of Energy Recovery, Membrane Rejection and Retentate Recycling, JMS, 2009, 339(1-2): 126--137.
4. Bartman, A., A. Zhu, P. D. Christofides and Y. Cohen, "Minimizing Energy Consumption in Reverse Osmosis Membrane Desalination Using Optimization-Based Control," Journal of Process Control, 20, 1261-1269, 2010.
5. Zhu, A., Christofides, P. D. and Cohen, Y. Energy Consumption Optimization of Reverse Osmosis Membrane Water Desalination Subject to Feed Salinity Fluctuation, IECR, 2009, 48 (21): 9581--9589.
6. Zhu, A., Rahardianto, A., Christofides, P. D. and Cohen, Y. Reverse Osmosis Desalination with High Permeability Membranes - Cost Optimization and Research Needs, Desalination and Water Treatment, 15(2010)256-266.
7. Zhu, A., P. D. Christofides and Y. Cohen, Effect of Stream Mixing on RO Energy Cost Minimization, Desalination, 261, 232-239, 2010.

See more of this Session: Membrane-Based Technologies for Water and Energy Reduction
See more of this Group/Topical: Topical 1: Water Technology for Developed and Developing Countries (see also Separations Division)

(465c) Hierarchical Nanoporous Filtration Membranes Enabled by a Roll-to-Roll Compatible Bottom-up Technology

Peng Jiang, Chemical Engineering, University of Florida, Gainesville, FL

Recent outbreaks of cholera in Haiti and swine flu over the world emphasize the importance in developing novel and inexpensive bioseparation media for removing pathogenic viruses and bacteria from drinking water and air. Macroporous membranes templated from self-assembled colloidal crystals consist of crystalline arrays of macro-sized voids which are interconnected by a network of monodispersed nano-sized pores. This unique hierarchical porous structure is promising for size-exclusive bioseparation. Unfortunately, traditional colloidal templating techniques suffer from low throughput and incompatibility with standard industrial manufacturing, which greatly hamper the development of practical macroporous membranes for challenging bioseparation. We have developed a novel roll-to-roll compatible doctor blade coating (DBC) technology that combines the simplicity and cost benefits of bottom-up self-assembly with the scalability and compatibility of standard industrial manufacturing. The platform enables the continuous production of large-area macroporous polymer membranes with highly uniform and interconnecting nanopores which have been successfully utilized for size-exclusive separation of pathogenic viruses and bacteria from water and air. The surface of the templated macroporous polymer membranes can be chemically functionalized to create self-cleaning and antifouling filters. The low manufacturing cost enabled by the DBC technology will allow it to be deployed in developing countries or disaster areas with high demand of clean water and air.

See more of this Session: Membrane-Based Technologies for Water and Energy Reduction
See more of this Group/Topical: Topical 1: Water Technology for Developed and Developing Countries (see also Separations Division)

(465d) Fundamental Water and Ion Transport Properties of Sulfonated Polysulfone

Joseph Cook, Chemical Engineering Program, University of Texas at Austin, Austin, TX, Chang Hyun Lee, Chemistry, Virginia Tech, Blacksburg, VA, Wei Xie, Chemical Engineering, The University of Texas at Austin, Austin, TX, Dr. Benny D. Freeman, Chemical Engineering, University of Texas, Austin, TX and James E. McGrath, Department of Chemistry, Virginia Polytechnic Institute and State University, Blacksburg, VA

This presentation will present results from an experimental study of the ion and water transport properties of bisphenol-A-based sulfonated polysulfone (BisA). Random copolymers were synthesized using 20-40% sulfonated monomer with non-sulfonated monomer, and water and ion solubility, diffusivity, and permeability coefficients were measured. In general, water uptake increased as the concentration of sulfonated monomer in these random copolymers increased. As water uptake increased, water permeability and salt transport increased. There was a general tradeoff observed, with polymers that were more permeable being less selective for water/salt separation. The transport properties (i.e., water and ion diffusion coefficients) were correlated with free volume in the hydrated polymers.

See more of this Session: Membrane-Based Technologies for Water and Energy Reduction
See more of this Group/Topical: Topical 1: Water Technology for Developed and Developing Countries (see also Separations Division)

(465e) Preparation of Robust Hybrid Silica Membranes for Reverse Osmosis

Toshinori Tsuru, Rong Xu, Jinhui Wang, Masakoto Kanezashi and Tomohisa Yoshioka, Department of Chemical Engineering, Hiroshima University, Higashi-Hiroshima, Japan

Membrane-based reverse osmosis (RO) is now the leading technology for production of pure water from seawater and brackish water. However, commercially dominant polyamide RO membranes suffer from poor tolerance to chlorine, which is the most widely used biocide in water treatment. Here, we developed a new type of hybrid silica membrane and were the first to apply this membrane to water purification by reverse osmosis. Sol-gel derived hybrid silica membranes were prepared using $(C_2H_5O)_3Si–CH_2CH_2–Si(OH_5C_2)_3$ (bis(triethoxysilyl) ethane, BTESE) in the design of the molecular sieve organic-inorganic silica network. Use of the organically bridging groups between two silicon atoms affords precise control of the pore size for the silica membranes according to the separation requirements. Our studies demonstrated that the hybrid silica membranes possessed an impressive hydrothermal stability, even under an elevated temperature of 90 °C. Furthermore, the prepared membranes exhibited an excellent chlorine resistance over a wide range of chlorine concentrations. After a total chlorine exposure of up to 35,000 ppm×h, there was no obvious change in membrane performance. We believe that the hybrid silica membranes with robust performance (high chlorine-resistance and excellent hydrothermal stability) will open a new realm of applications of reverse osmosis.

See more of this Session: Membrane-Based Technologies for Water and Energy Reduction
See more of this Group/Topical: Topical 1: Water Technology for Developed and Developing Countries (see also Separations Division)

INVESTIGATION OF PROTEIN ADSORPTION BEHAVIOR ON POLYMER SURFACES USING ATOMIC FORCE MICROSCOPY AND QUARTZ CRYSTAL MICROBALANCE WITH DISSIPATION

Toru Ishigami, Keisuke Hirami, Dept. of Chemical Science and Engineering, Center for Membrane and Film Technology, Kobe University, Kobe, Hyogo, Japan
Masatoshi Hashino, Noboru Kubota, Asahi Kasei Chemicals Coorporation, Fuji, Shizuoka, Japan
Yoshikage Ohmukai, Eiji Kamio, Tatsuo Maruyama, Hideto Matsuyama, Dept. of Chemical Science and Engineering, Center for Membrane and Film Technology, Kobe University, Kobe, Hyogo, Japan

Introduction

Membrane fouling is the most serious problems in the water treatment with membrane. Previous studies have shown that the hydrophobic interaction between foulants such as proteins and polymeric membranes used for the water treatment causes the serious membrane fouling. However, the mechanism has not been well known. In this study, the fouling behavior, the adsorption behavior and the interaction between protein and polymers were focused on.

Experimental

Bovine serum albumin (BSA) was used as a foulant and polyvinylidene difluoride (PVDF), polyethylene-polyvinyl alcohol copolymer (EVOH) and polyether sulphone (PES) were used as polymers. To grasp the fouling behavior, hollow fiber membranes from these polymers with the similar pore sizes and water permeabilities were prepared through non-solvent induced phase separation (NIPS) method. After that, the filtration and the backwashing experiments were performed with these membranes and BSA. In addition, the adsorption behavior on the polymer surface and the interaction forces between the foulants and the polymers were investigated using quartz crystal microbalance with dissipation (QCM-D) and atomic force microscopy (AFM), respectively.

Result and Discussion

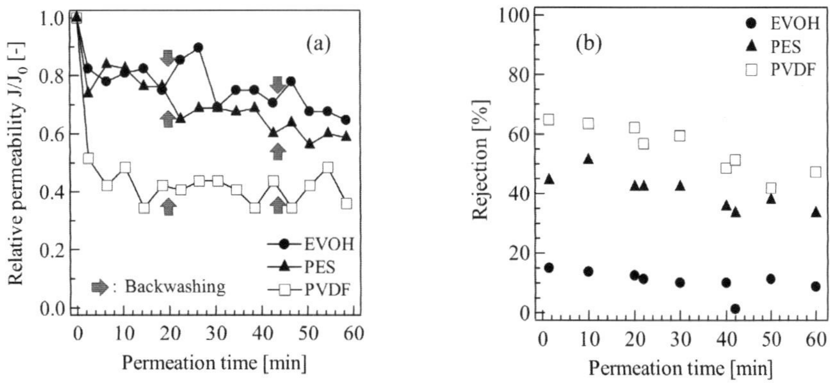

Fig. 1 Time variation of relative permeability (a) and BSA rejection (b).

Fig. 1 shows the time variation of the relative permeability and the BSA rejection, obtained in the filtration experiment with BSA solution. From this, it was found that the water permeabilities decreased as the time passed. The permeability decrease was the highest in the case of PVDF and was the lowest in the case of EVOH. In addition, although backwashing operation was performed, the obvious recoveries of the water permeabilities was not observed.

Fig. 2 shows the adsorption isotherms for BSA on polymer surfaces obtained from QCM-D experiment, the amount of adsorption becomes higher in the order of PVDF, PES and EVOH. This tendency was the same as the result of the filtration experiment. In addition, it was found that BSA adsorbed on any polymer surfaces with a single layer, because the plots corresponded with Langmuir isotherm.

Fig. 3 shows the relationship between the frequency shift, Δf, and the dissipation shift, ΔD, obtained in QCM-D measurement. It was found that BSA formed the rigid layer at initial adsorption in the case of PVDF. This is because BSA deforms on the polymer surface by the strong hydrophobic interaction between PVDF and BSA. In contrast, BSA was found to adsorb on both PES and EVOH surface in a soft layer.

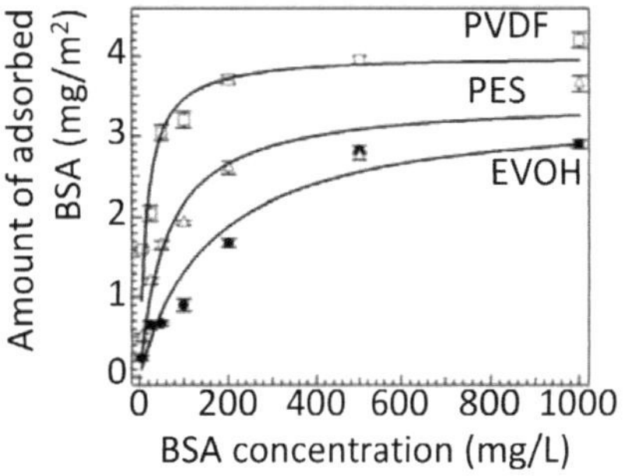

Fig. 2 Adsorption isotherms for BSA on polymer surfaces.

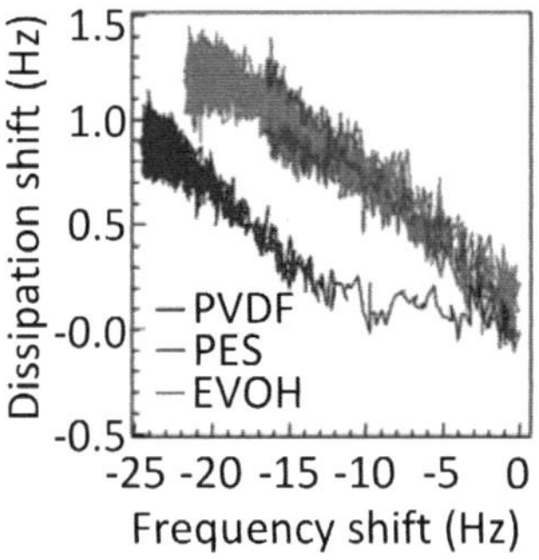

Fig. 3 Dissipation shift against frequency shift for BSA adsorption on polymer surfaces.

Next, the adsorbed BSA layer with an adsorption equilibrium was attempted to desorbed by shear flow of buffer solution. Table 1 shows the ratio of the desorption amount to the adsorption amount. The desorption ratio were very small in each polymer and in each BSA concentration. Thus, it is found that it is difficult to separate the adsorbed BSA from polymer surfaces. This result supports that the backwashing is not effective in the filtration experiment, as described above.

Table 1. Ratio of the desorption amount to the adsorption amount.

	50 mg·L^{-1} of BSA solution	**1000 mg·L^{-1} of BSA solution**
EVOH	5.9%	2.0%
PES	4.9%	2.4%
PVDF	3.8%	2.4%

Fig. 4 shows the adhesion forces between polymers and BSA, obtained in AFM measurement. It was found that the adhesion forces from any polymer surfaces were almost the same. In addition, the adhesion forces between BSA and BSA on surface were much lower compared with those between BSA and polymer surfaces. Thus, this result supports that BSA adsorbed in monolayer, obtained in QCM-D.

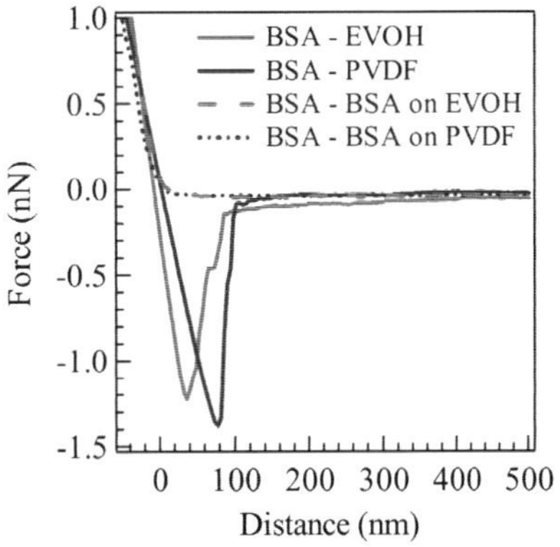

Fig. 4 Adhesion forces of BSA colloid probe to polymer surfaces and those covered with BSA

Conclusion

To predict the membrane fouling, we investigated the affinity between BSA and polymer surface. From the result of fouling experiment, in the case of PVDF hollow fiber membrane, the severe fouling occurred and the backwashing was not effective to the recovery of the water permeability. Based on AFM and QCM-D measurement, BSA easily adsorbed and was able to undergo a conformational change onto PVDF. In addition, BSA was hardly separated from any polymer surfaces. In contrast, the interaction between BSA and BSA was very weak. Thus, BSA adsorbed with monolayer onto any polymer surfaces and this result corresponded with that obtained in QCM-D.

(545a) Improving Electrospun Nanofiber Strength for Filtration Applications

Liwei Huang[1], Ngoc Bui[1], Jason Arena[1], Seetha Manickam[1] and **Jeffrey R. McCutcheon**[2], (1) Chemical, Materials & Biomolecular Engineering, University of Connecticut, Storrs, CT, (2)Department of Chemical, Materials and Biomolecular Engineering, University of Connecticut, Storrs, CT

Electrospun nanofibers have recently been considered for water filtration. These nano-scale fibers have a high specific surface area combined with a high permeability, thus enabling their use in low pressure microfiltration and adsorption based separations. However, electrospun nanofibers are generally characterized as having weak mechanical properties. As such, they are typically attached to a secondary support, such as a polyester nonwoven. The use of non-woven fabric might be unfavorable because it could increases the filtration resistance, especially in low pressure filtration.

In this study we present several methods for increasing the strength of standalone electrospun nanofibers. Our initial polymers in this study include polysulfone and polyacrylonitrile, two polymers commonly used in filtration applications. Through tuning of the spinning parameters (humidity and solvent), we were able to control the tensile strength and modulus of the nanofiber mats. Post treatment of the nanofibers also resulted in increased strength. Fiber size, surface area, morphology and pore size/ structure were also characterized during this study.

See more of this Session: Novel Materials Development In Water Purification
See more of this Group/Topical: Topical 1: Water Technology for Developed and Developing Countries (see also Separations Division)

(545b) Surface Modification of Membranes to Improve Fouling Resistance

Sirirat Kasemset[1], Albert Lee[1], Daniel J. Miller[1], Benny D. Freeman[1] and Mukul M. Sharma[2], (1) Chemical Engineering, University of Texas at Austin, Austin, TX, (2)Department of Petroleum & Geosystems Engineering, University of Texas at Austin, Austin, TX

Membrane technology is an important approach for water purification. One of the major problems of this technology is membrane fouling, which causes a decline in permeate flux and increase in transmembrane pressure drop. Membrane surface modification is one technique to improve fouling resistance of membranes. A bio-inspired material called "Polydopamine (PDOPA)" has been used to modify membrane surfaces and improve membrane fouling resistance based upon significantly enhanced permeate flux from oil-water emulsion filtration. This presentation focuses on studies to understand the modification mechanism and fundamental basis for improved fouling properties of membranes coated with PDOPA.

In this study, three surface modification protocol parameters have been studied using reverse osmosis membranes. The parameters are dopamine coating solution concentration, dopamine deposition time, and buffer solution pH. Constant transmembrane pressure difference crossflow filtration fouling studies were used to probe the influence of these parameters on fouling performance.

See more of this Session: Novel Materials Development In Water Purification
See more of this Group/Topical: Topical 1: Water Technology for Developed and Developing Countries (see also Separations Division)

(545c) Naturally-Functionalized Antimicrobial Sand (f-sand) Using Moringa Oleifera Seeds

Darrell Velegol[1], Huda A. Jerri[1], Kristin J. Adolfsen[1], Lauren R. McCullough[1] and Stephanie B. Velegol[2], (1)Chemical Engineering, Penn State University, University Park, PA, (2)Civil Engineering, Penn State University, University Park, PA

Moringa oleifera (Moringa) seeds contain a natural cationic protein (MOCP), which has been used as an antimicrobial flocculant for water clarification. The main challenge with using Moringa seeds for this purpose is that they release other water-soluble proteins and organic matter, which increases the concentration of dissolved organic matter (DOM) in the water. The presence of this DOM supports the re-growth of pathogens in treated water, preventing its storage and later use. We have established a strategy for retaining the MOCP, while removing the excess BOD. First, the MOCP is electrostatically immobilized onto sand grains and then the excess BOD is rinsed away, inhibiting the growth of bacteria in the purified water. The data indicate that the functionalized sand (f-sand) removes microparticles and pathogens from water, renders adhered E. coli bacteria non-viable, and significantly reduces turbidity of a kaolin suspension. The surface charge and antimicrobial properties of f-sand were ascertained with confocal microscopy using fluorescent polymer nanoparticle and microparticle tagging and fluorescent (live-dead) staining of bacteria. The reduction of DOM, which can contribute to bacterial re-growth, was demonstrated by a considerable reduction in the solution biochemical oxygen demand (BOD). These results suggest that immobilization of MOCP onto sand can provide a simple, locally-sustainable process for producing storable drinking water.

See more of this Session: Novel Materials Development In Water Purification
See more of this Group/Topical: Topical 1: Water Technology for Developed and Developing Countries (see also Separations Division)

(545e) Nano-Pipette Directed Motion of Functionalized Transmembrane Channel

Meenakshi Dutt[1], Olga Kuksenok[2] and Anna C. Balazs[2], (1)Chemical and Petroleum Engineering, University of Pittsburgh, Pittsburgh, PA, (2)Chemical Engineering, University of Pittsburgh, Pittsburgh, PA

Via the Dissipative Particle Dynamics approach, we design a system that allows transport of a nanochannel to a desired location by applying an external force. Each nanochannel encompasses an ABA architecture, with a hydrophobic shaft (B) with two hydrophilic ends (A). One of the hydrophilic ends of the nanochannel is functionalized with hydrophilic functional groups, or hairs. The hydrophilic hairs serve a dual role: (1) control transport across the membrane barrier when the channel diffuses freely in the membrane, and (2) enable the channel relocation to a specific membrane site. Our system comprises a transmembrane hairy nanochannel with the hairs extending into solution. In our earlier work, we demonstrated the spontaneous insertion of such a hairy nanochannel into a lipid bilayer (Nanoscale, 2011, 3, 240). First, we hold a suitably functionalized pipette stationary above the membrane while the nanochannel freely diffuses within the membrane. For an optimal range of parameters, we demonstrate that the hairs find the pipette and spontaneously anchor onto it. We then show that by moving the pipette for a range of velocities, we can effectively transport the channel to any location within the membrane. This prototype system can provide guidelines for designing a number of biomimetic applications.

See more of this Session: Novel Materials Development In Water Purification
See more of this Group/Topical: Topical 1: Water Technology for Developed and Developing Countries (see also Separations Division)

(575a) Treatment of Produced Water by Ceramic Membranes

Bruce A. Bishop[1], Stanton Smith[1] and Lnsp Nagghappan[2], (1)HPD Membrane Technology Center, Veolia Water Solutions and Technologies, Waltham, MA, (2)North American Water Systems, Veolia Water Solutions and Technologies, Moon Township, PA

Ceramic membranes are at the interface of energy and waterthrough their application to produced water treatment. In the production of oil and gas, water fromthe reservoir is brought to the surface and this is termed produced water. Worldwide, the ratio of water to oil thatcomes from the wells is about 3:1, with the US having a much higher ratio thanthe worldwide average. The amount ofproduced water generated each year is astounding. Most produced water is reinjectedinto disposal wells but given the potential recycle and reuse applications,there is a strong need to treat the various types of produced wateravailable. CeraMem" ceramic membranes, aloneand in combination with other unit operations, are being used to treat producedwater of different types. The marketdrivers for and results of produced water treatment processes using CeraMem" ceramic membranes will bereviewed.

See more of this Session: Ceramic and Mixed Matrix Membranes for Water Treatment and Reuse
See more of this Group/Topical: Topical 1: Water Technology for Developed and Developing Countries (see also Separations Division)

(575b) Electrosteric Stabilization and Covalent Functionalization of Iron Nanoparticles In a Mixed Matrix Membrane

Lauren F. Greenlee and Jessica D. Torrey, Materials Reliability Division, National Institute of Standards and Technology, Boulder, CO

There has been a growing interest in using iron-based nanoparticles in water treatment due to their catalytic properties. Monometallic and bimetallic iron nanoparticles are reactive in an aqueous system and can be designed for targeted removal of specific organic and inorganic water contaminants. Iron metal nanoparticles undergo surface oxidation through reactions with both water and oxygen and produce reactive species that are then utilized to oxidize or reduce a specific contaminant. Through the oxidation process, iron metal is oxidized to ferrous iron, which in turn can participate in the Fenton reactions and produce reactive oxygen species and hydroxyl radicals. Iron metal oxidation also produces hydrogen gas (through reaction with water), which can react with organic compounds through reduction reactions. The catalytic properties of iron nanoparticles can be enhanced through particle stabilization by organic chelating compounds and through secondary metal coatings. Both organic stabilizers and secondary metals (*e.g.*, palladium, copper) affect the oxidation of the nanoparticles and ultimately, nanoparticle reactivity.

For reactive nanoparticles to be used in water treatment, the particles must be contained or immobilized within the treatment system. One potential method to contain the nanoparticles is to incorporate the particles into polymeric membranes; ideally, these mixed matrix membranes would be designed such that the membrane continues to function as a physical barrier to filtrants, while the reactive nanoparticles degrade or adsorb dissolved contaminants that typically pass through conventional membranes.

This research investigates the use of a bifunctional poly(ethylene glycol)-based polymer in iron nanoparticle functionalization and incorporation into a poly(ether sulfone) (PES) film. The effect of covalent attachment of the polymer matrix to the iron nanoparticle on separation ability will be evaluated as a function of filtration performance in a dead-end permeation cell. This mixed matrix film will be compared to a synthesized PES film where the iron nanoparticles are contained within the film by electrosteric stabilization. Mixed matrix films will be characterized by Fourier transform infrared spectroscopy (FTIR), atomic force microscopy (AFM), and electron microscopy techniques. The comparison between covalently-attached nanoparticles and nanoparticles sterically and electrostatically bound to the polymer matrix will provide insight as to how nanoparticle stabilization within a polymeric membrane affects membrane performance and nanoparticle dispersion within the film.

See more of this Session: Ceramic and Mixed Matrix Membranes for Water Treatment and Reuse
See more of this Group/Topical: Topical 1: Water Technology for Developed and Developing Countries

(575c) TiO2-Enhanced Polymer-Derived Ceramic Composite Membranes for Water Purification

Jessica D. Torrey and Lauren F. Greenlee, Materials Reliability Division, National Institute of Standards and Technology, Boulder, CO

TiO_2 nanoparticles have shown potential in the removal of volatile organic compounds and pharmaceuticals from wastewater by heterogeneous photocatalysis. While the photooxidative effect of TiO_2 nanoparticles under UV light has been widely studied and seen many improvements, the issue of immobilization of the nanoparticle catalyst remains an obstacle to the realization of widespread practical application. Additionally, components of the degradation reaction- namely the UV source itself and the highly reactive species produced in water- cause oxidation of the matrix when TiO_2 is immobilized in an organic membrane. In this research, we utilize silicon-based preceramic polymer as a novel matrix material to produce a titanium dioxide-enhanced inorganic polymer-derived ceramic composite membrane. The influence of the addition of TiO_2 on the morphology of the membrane will be presented, including porosity, thickness, and nanoparticle dispersion. Properties, including contact angle, membrane permeability, and methylene blue degradation, of the unloaded and nanoparticle-loaded membranes will be compared. Quartz crystal microbalance analysis will show the stability of the film under ultraviolet irradiation.

See more of this Session: Ceramic and Mixed Matrix Membranes for Water Treatment and Reuse
See more of this Group/Topical: Topical 1: Water Technology for Developed and Developing Countries (see also Separations Division)

(575d) Silica-Polymer Mixed Matrix Membranes for Mercury and Silver Sorption

Noah D. Meeks, Dept of Chemical and Materials Engineering, University of Kentucky, Lexington, KY and Dibakar Bhattacharyya, Chemical & Materials Engineering, University of Kentucky, Lexington, KY

Mercury contaminated waters, from lakes and rivers to mine runoff and other contaminated streams, poses a serious public health hazard. Similarly, silver sorption is often desired as a cost-recovery mechanism for various industries, including mining. Novel functionalized silica particles have been synthesized with a Hg(II) sorption capacity of up to 85 mg/g (200 ppm Hg feed concentration at pH ~ 5.5). However, like many sorbent particles, use of these materials in a packed bed or slurry has some drawbacks, such as internal mass transfer limitations as well as handling of hazardous and/or potentially valuable microparticles. Novel functionalization techniques allow for mechanically stable mixed matrix membranes (MMM). These are synthesized with higher loadings of sorbent (up to 50 wt% sorbent) than has been reported for previous silica/polymer MMM (30 wt% sorbent). Functionalized silica/polymer MMM offer advantages for the efficient sorption of these ions onto the novel silica particles. Synthesis of these MMM disperses the sorbent particles into a polymer matrix, so that the active sites are more exposed. Convective flow used in the membrane processes exploits this dispersion for increased mass transfer and access to the active sorption sites. Finally supporting the sorbent particles within a membrane matrix allows for facile handling, whether for disposal or adsorbate recovery. This work is supported by NEI Corporation.

See more of this Session: Ceramic and Mixed Matrix Membranes for Water Treatment and Reuse
See more of this Group/Topical: Topical 1: Water Technology for Developed and Developing Countries (see also Separations Division)

(594a) Characterization of Pecan Shell Activated Carbon In Aqueous Chromium Systems

Venkata Daram, Chemical Engineering, New Mexico State University, Las Cruces, NM

CHARACTERIZATION OF PECAN SHELL ACTIVATED CARBON IN AQEOUS CHROMIUM SYSTEMS

Venkata Daram**,** *New Mexico State University, Las Cruces, New Mexico-88003*

Chromium (VI) belongs to the family of heavy metals and is considered a highly toxic pollutant in wastewater streams. Chromium and its compounds are widely used in leather tanning, electro-plating, anodizing baths, and rinse waters. Pecan shell, a waste product of the agricultural industry, was used for the removal of chromium (VI) from wastewater streams. Activated carbon was produced by chemical activation of pecan shells with phosphoric acid followed by oxidation with air according to the method described by Dr. Rockstraw and Dr. Syed Dastgheib. Equilibrium studies were carried out at constant pH with carbonization temperature serving as the independent study variable. Characterization studies of produced activated carbon were carried out using nitrogen adsorption isotherms, and BET surface areas were determined and found to be 555, 822, 1547, 1432 m^2/g for carbons carbonized at 360, 410, 450 and 500°C respectively. The equilibrium isotherm data was fit using Langmuir and Freundlich isotherms, and with the latter providing a better fit of the chromium (VI) equilibrium adsorption data. The maximum adsorption capacity for chromium (VI) based on the Freundlich model was found to be 1.92 mmol/g Cr (VI).

See more of this Session: Low Cost Adsorbents for Water Purification
See more of this Group/Topical: Topical 1: Water Technology for Developed and Developing Countries (see also Separations Division)

BIOSORPTION OF LEAD BY SPENT COFFEE GROUND: KINETIC AND ISOTHERM STUDIES

N.E. Dávila-Guzmán[a], F.J. Cerino-Córdova[a], J. R. Rangel-Méndez[b], P.E. Díaz-Flores[c]
[a]Universidad Autónoma de Nuevo León, 66451, S.N. de los G., N.L., México.
[b]Instituto Potosino de Investigación Científica y Tecnológica, S.L.P., S.L.P, México.
[c]Academic Coordination Altiplano Region, Universidad Autónoma de San Luis Potosí, Matehuala, S.L.P., México

ABSTRACT

The use of spent coffee ground (SCG) for the removal of Pb (II) from aqueous solution has been investigated as a potential low-cost treatment method for heavy metal adsorption. Batch experiments were performed to obtain kinetics and adsorption isotherm. The heavy metal adsorption capacities of SCG were studied as a function of adsorbate concentration (5, 120 and 207 mg/L) and stirring speed (200, 300 and 400 min^{-1}). Equilibrium isotherms for the adsorption of Pb (II) were analyzed by the Langmuir and Freundlich isotherm models. The Langmuir isotherm model was found to represent better the experimental data (%ε= 4.00). The maximum removal of Pb (II) is found to be 40 mg/g at pH 4.5 and initial Pb (II) concentration of 207 mg/L. The effect of stirring speed was found to a significant factor for the external mass transfer resistance at 200 min^{-1} and the external mass transfer coefficients were estimated by the Mathews and Weber model. The Pseudo First Order model and Crank model were used to fit the kinetic data, the results suggest that intraparticle diffusion could be the rate controlling step with an intraparticular diffusion coefficient 1.46×10^{-11} m^2/s. The results indicate that the SCG could be considered as a low cost potential adsorbent for Pb (II) removal from aqueous solutions.

INTRODUCTION

Water pollution by heavy metal is a serious global environmental problem [1]. In Mexico, 1×10^{12} m^3 of wastewater were discharged from industry in 2009 [2], such discharges can contain several heavy metals. These pollutants can cause adverse effects on human health, such as irritable bowel, muscular weakness, even blindness, amnesia and can even cause death [3]. Several techniques have been used to remove heavy metals, such as, chemical precipitation, electrodialysis, ultrafiltration, ionic exchange, adsorption and biosorption [4-9]. The latter has become popular over the last decades because the biosorbents used are economical, environmentally friendly and exhibit some selectivity for heavy metals [10]. Agro-industrial wastes, such as tea, rice husk and coffee bean, have been studied for the heavy metals removal. The capacity of these materials to remove pollutants can be due to organic compounds (polysaccharides, flavonoids, alkaloids etc.), present in their structures [11-12]. Coffee is one of the major agricultural products used in beverage industry, being Brazil the largest coffee producer in the world while Mexico ranks fifth with an estimated production of 2.58×10^5 tons of coffee in 2007 [13]. This suggests a high availability of solid coffee waste.

In coffee producing countries, residues are a source of serious pollution and environmental problems, and several investigations has been conducted for the revaluation of the coffee industry waste, such as the use of coffee pulp for the cultivation of edible fungi of the genus *Pleurotus*, pulp processing into compost using red worms, production of food yeast, among others [14-17]. Coffee residues have been used in water treatment, in removal of heavy metals (Cd (II), Cu (II), Pb (II) and Zn (II) in particular) present in industrial effluents [18-19]. However, studies in this area are very recent, and there is not information about the adsorption kinetics process and the rate controlling step, which are important for design purpose of adsorption system. This work investigated the feasibility to use spent coffee ground as adsorbent for the removal of Pb (II) from aqueous solution. In this study, equilibrium and kinetic studies were realized with the modeling of the isotherm and the adsorption kinetics for find out the parameters associated to these process.

MATERIALS AND METHODS

Biosorbent and Metal Solution.

Spent coffee ground (SCG) was collected locally and sieved through a #35 ASTM sieve. SCG was pretreated with 0.5M NaOH solution at 40°C and 100 min^{-1} during 4h .The solid/liquid ratio used in the pretreatment was 100 g/L. After, the SCG was washed repeatedly with deionized water until pH close to 6.0 was achieved. Finally, the SCG was dried at 60°C for 60h and cooled to room temperature. All reagents used in this study were analytical-reagent grade. The Pb (II) solution was prepared by dissolving appropriate amounts of $Pb(NO_3)_2$.

Adsorption Batch Experiments.

An adsorber system in a standard batch stirred tank configuration was used in the experiments [20]. The absorber system (Figure 1) was a 1.2 L glass vessel (1), 10 cm in diameter and 15 cm high, which was immersed into a bath temperature at 25°C. Six evenly spaced vertical baffles (2) 1 cm wide were fitted on the wall of the adsorber in order to prevent the formation of a vortex. When the stirrer rotates, the centrifugal action produces a rapid circulating flow of solution entering the cage (3) at the bottom and leaving through the radial holes of the caging, allowing a vigorous hydrodynamic condition surrounding the SCG, which reduces the diffusional resistances in the liquid film around SCG [21]. A fixed mass of SCG (0.75g) was loaded within the cell of the stirrer for conditioning with 0.75 L acidified water (pH 4.5) at 400 min^{-1} for 90 min. After, the stirrer (4) was turned off and the acidified water was replacement for new acidified water (0.5 L) at pH 4.5, then 0.25 L of concentrated metal solution was adding to obtain a desirable concentration value after dilution to 0.75 L, rapidly the stirrer was turned on at fixed speed. The experiments were performed during 80 min. The pH of the solutions was measured (5) and kept at pH 4.5 during the adsorption test using 0.1 M NaOH or HNO_3 solution. Samples were collected at different times, and the metal concentrations were analyzed by atomic absorption spectrometry (Thermo Scientific®). Adsorption isotherm study was carried out with different initial concentration of metal ion from 5-207 mg/L at 400 min^{-1}, 25°C, pH 4.5 and 0.750 g of SCG. The amount of Pb (II) adsorbed at equilibrium, q_e, was calculated as follows:

$$q_e = V(C_o - C_e)/m \qquad (1)$$

Where C_o and C_e are the Pb (II) initial and at equilibrium concentrations (mg/L), respectively; V is the solution volume in L; m is the dry weight of SCG in g; q is the uptake capacity at the time t in mg/g. The stirring speed effect is studied for purpose of minimize the external mass transfer resistance, the initial concentration of heavy metal solutions was 207 mg/L and the stirring speed was varying from 200-400 min^{-1}. The effect of adsorbate concentration is studied by varying the initial metal concentration (5, 120 and 207 mg/L) with a stirring speed of 400 min^{-1}, pH 4.5 and 25°C.

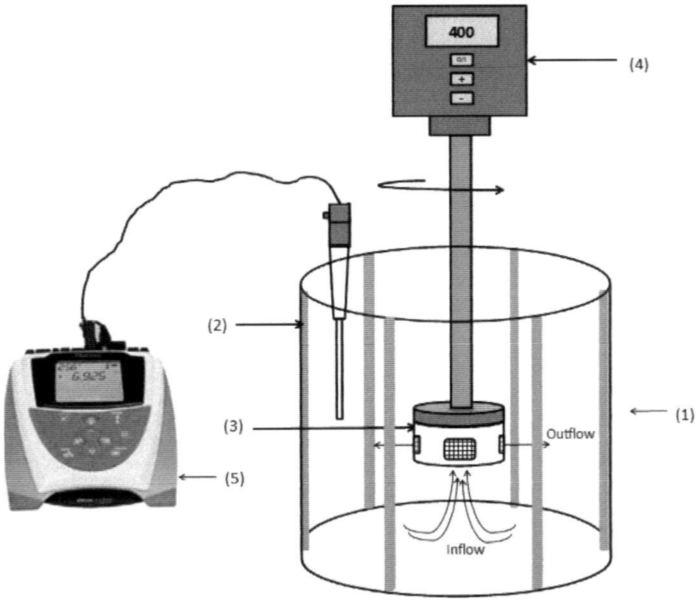

Figure 1. Standard Tank Configuration for Adsorption Studies.

Nonlinear Regression Analysis.

All the kinetics and isotherms model parameters were evaluated by nonlinear regression minimizing the error function of average error relative (ARE) by the Solver tool of Microsoft Excel. ARE was calculated from the Eq. (2):

$$\%\varepsilon = \frac{\sum_{i=1}^{n} \left| \frac{q_{exp_i} - q_{sim_i}}{q_{exp_i}} \right|}{n} \times 100\% \qquad (2)$$

RESULTS AND DISCUSSION

Isotherm Models.

Figure 2 shows the adsorption isotherm, i.e., the relations between the amount of Pb (II) adsorbed per unit mass of SCG (q_e) and their remaining concentrations in the aqueous solutions (C_e). The Langmuir and Freundlich isotherm models were used in this work to fit experimental data. They differ in the basic assumptions, shape of the isotherm and nature of the adsorbent surface [22-23]. The equations and model parameters are shown in Table 1. The Langmuir model represent well the Pb (II) adsorption based on the ARE ($\%\varepsilon = 4.00$), while the Freundlich model shows a more deviation from the experimental data ($\%\varepsilon = 8.77$). The maximum adsorption capacity (q_m) obtained in this study was 35.27 mg/g. Utomo and Hunter [24] have reported a maximum adsorption capacity of 49.47 mg/g for the adsorption of Pb (II) onto exhausted coffee ground, which is 40% more than that obtained in the present work; however, they used a concentration adsorbent three times larger than that used in this study, which could be explain the enhance in the maximum adsorption capacity.

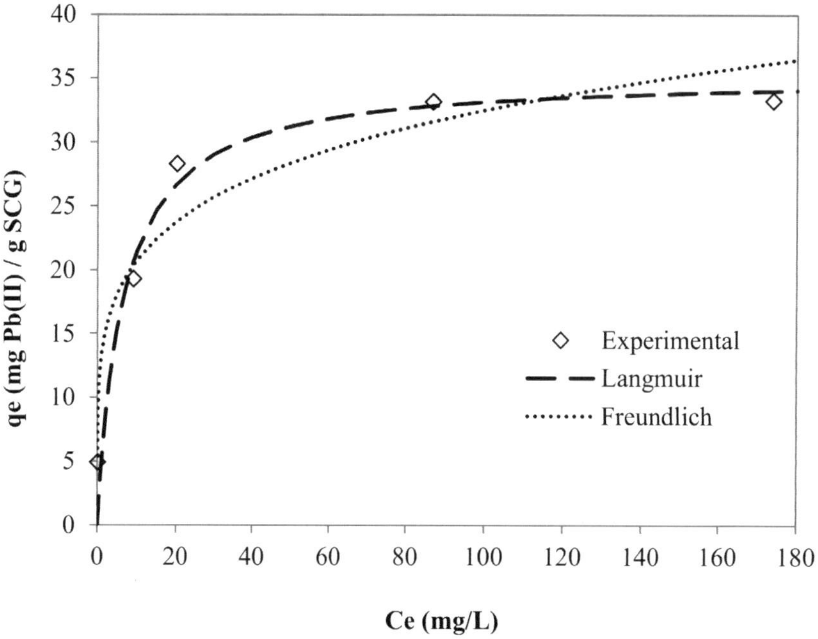

Figure 2. Isotherm Models of Pb (II) Adsorption onto SCG at a pH 4.5 and 25°C.

Table 1. Parameters and Average Relative Errors for Adsorption Isotherm Models.

Model	Equation	Parameters	Value	ε (%)
Langmuir	$q_e = \frac{q_m b C_e}{1 + bC_e}$	q_m: Maximum adsorption capacity (mg/g) b: Langmuir parameter (L/mg)	35.27 0.155	4.00
Freundlich	$q_e = K_F C_e^{1/n_F}$	K_F: Freundlich parameter (mg/g)(L/mg)$^{1/nF}$ n: Freundlich parameter	13.24 5.13	8.77

Effect of Adsorbate Concentration.

The effect of initial metal concentration is important when designing adsorption treatment systems since a given mass of adsorbent can only adsorb a fixed amount of adsorbate [25]. Experimental data are shown in Figure 3 for the adsorption of Pb (II) onto SCG at different initial metal concentration (5, 120 and 207 mg/L). An increase in the initial Pb (II) concentration leads to an increase in the adsorption capacity and the equilibrium uptake. This is a result of the increase in the driving force of the concentration gradient, with an increase in the metal ion initial concentration. The equilibrium time required for the metal ion is short because the majority of Pb (II) is removed within the first 20 min., for the low initial lead concentration (5 mg/L) and within the first 50 min. for the medium and high initial lead concentration (120 and 207 mg/L). The uptake of Pb (II) was found to be more rapid from solutions with lower concentrations which could be attributed at the less competition between the Pb (II) ions for available surface for adsorption.

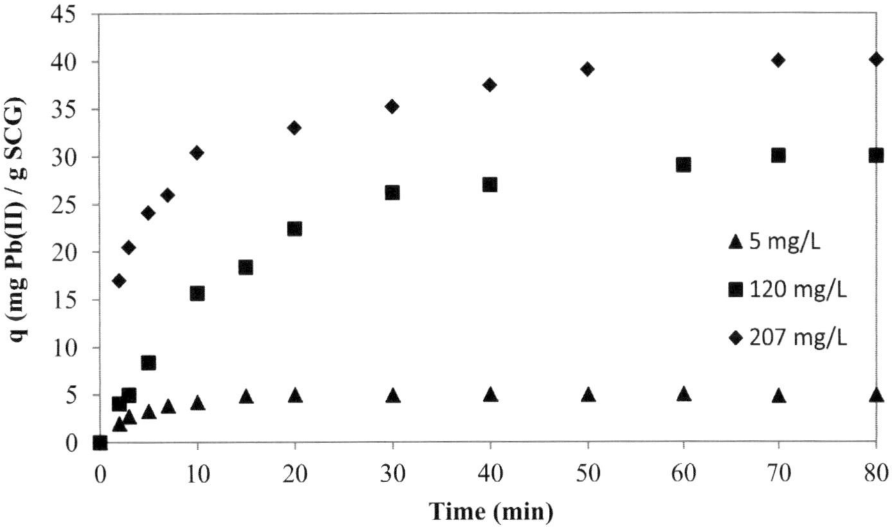

Figure 3. Effect of initial concentration of Pb (II) on the adsorption capacity. Conditions: pH 4.5, stirring speed= 400 rpm, temperature= 25°C, adsorbent dosage = 1 g/L.

Effect of Stirring Speed.

The decay curves of Pb (II) concentration at different stirring speed were carried out in order to find out the conditions that minimizes the resistance of external mass transfer. Figure 4 shows that the decay of Pb (II) concentration is the same for the stirring speed of 300 and 400 min^{-1}, which indicate that the mixing is strong enough for reducing the external mass transfer. Moreover, the decay curve at 200 min^{-1} is slightly less sharp than the decay curves at 300 and 400 min^{-1}, because the diffusion rate of metal ion from the bulk liquid to the liquid boundary layer surrounding adsorbent particles becomes lower due to a reduction of turbulence which not allows the narrowing of the thickness of the liquid boundary layer. To determine the extent of external mass transfer resistance around SCG, a simple approach has been proposed for estimating the mass transfer coefficient in the boundary layer during the biosorption process using the Mathews and Weber (M&W) model. The M&W model is an application of the Fick's law [26] which expresses the evolution of solute concentration in the solution by the following equation:

$$\frac{C_t}{C_0} = e^{-K_f S t} \tag{3}$$

Where k_f is the external mass transfer coefficient (cm/min); C_o and C_t are the copper concentration at t=0 and t=t, respectively; S is the external surface of adsorbent per unit volume of solution (cm^{-1}), which can be obtained by the following equation:

$$S = \frac{6m_s}{d_p \rho_t (1 - \varepsilon_p)} \tag{4}$$

Where m_s (0.001 g/cm^3) is the adsorbent mass per unit volume of adsorbate, d_p (0.0603 cm) is the particle diameter, ρ_t is the particle density (0.747g/cm^3) y ε_p (0.474) is the particle porosity. The particle density and porosity belong to roasted coffee bean, because there is not information about ground coffee, and this data were taken as approximation of textural properties for SCG [27]. The models of M&W represent adequately the experimental data for the first 10 min. of biosorption (when the external mass transfer resistance is more meaningful) with ARE less than 5%. The coefficients of external mass transfer for stirring speeds of 300 and 400 min^{-1} obtained with M&W model was found to be close to each other, nevertheless at 200 min^{-1} the k_f was lower than the value obtained at 300 and 400 min^{-1}, indicating that external mass transfer rate was reduced with the increase in the stirring speed (Table 2).

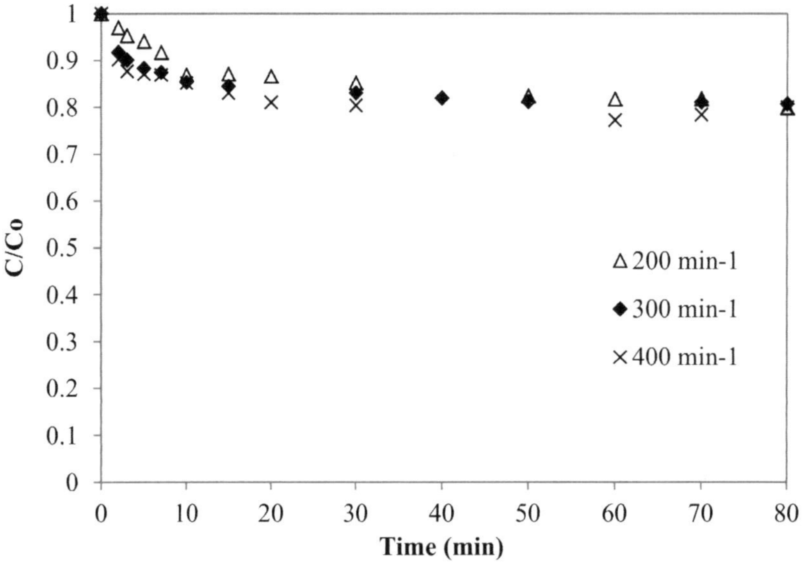

Figure 4. Effect of stirring speed in the decay curve of Pb (II) concentration during the adsorption onto SCG. Conditions: pH 4.5, C_o= 207 mg/L, temperature= 25°C, adsorbent dosage = 1 g/L.

Table 2. Mass Transfer Coefficient Estimated by M&W Model.

Stirring speed (min^{-1})	K_f (m/s)	ε (%)
200	8.33×10^{-6}	0.55
300	1.31×10^{-6}	3.50
400	1.31×10^{-5}	2.70

Kinetic Modeling.

The performance and ultimate cost of an adsorption system depends on the effectiveness of the process design and the efficiency of process operation which requires an understanding of the kinetics of uptake and the identification of the rate-determining step [28]. The adsorption process in a porous material can be described in terms of the following three steps: (1) a mass transfer of solute from the bulk solution to the particle surface through the film (film diffusion), (2) diffusion of the solute within the adsorbent particle to the adsorption sites (intraparticle diffusion) and (3) final uptake of solute at the adsorption sites [29]. Kinetic models available in the literature have been applied to the experimental data obtained to investigate the mass transfer mechanism and to get the potential rate-controlling step. According to the results obtained from the effect of stirring speed, the external mass transfer did not play a significant role in the kinetic adsorption, due to this, external mass transfer models were not used in this work. The Crank model, and Pseudo First Order model (PS-1) were used for modeling the kinetic adsorption of Pb (II) onto SCG. The Crank model assumes that the external film resistance is negligible and only takes account the intraparticular diffusion, and the PS-1 describes the adsorption rate based on the adsorption capacity, neglecting the film and intraparticular diffusion [30-31]. The kinetic parameters and the average relative errors are reported in Table 3. The PS-1 model did not represent adequately the experimental data according to ARE ($\%\varepsilon = 9.00$) due to limitations by neglecting the mass transfer resistances. Moreover, the Crank model presented a satisfactory fit ($\%\varepsilon = 3.00$) and an intraparticle diffusion coefficient with a value of 1.46×10^{-11} m^2/s. This means, according to Crank's model, that the external mass transfer is negligible, suggesting that the adsorption of Pb (II) onto SCG is controlled by intraparticle diffusion.

Table 3. Parameters and Average Relative Errors for Adsorption Kinetic Models.

Model	Equation	Parameters	Value	ε (%)
Crank	$\frac{q_t}{q_e} = 1 - \frac{6}{\pi^2}\sum_{n=1}^{\infty}\frac{1}{n^2}\exp\left(-\frac{D_s n^2 \pi^2 t}{R^2}\right)$	Ds: Intraparticle diffusion coefficient (m^2/s)	1.46×10^{-11}	3.00
PS-1	$q = q_e\left(1 - e^{-kt}\right)$	K_1: Rate constant (min^{-1})	0.184	9.00

CONCLUSIONS

The spent coffee ground is a suitable adsorbent for the removal of Pb (II) form aqueous solutions. The adsorption of Pb (II) is highly dependent on the initial metal ion concentration. The maximum uptake adsorption was 40 mg/g at initial metal concentration of 207.19 mg/L. The isotherm adsorption follows a Langmuir isotherm model. The studies of the effect of stirring speed show that increase in the stirring speed decrease the external mass transfer resistance. The adsorption kinetic studies show that intraparticle diffusion is the rate controlling step following an intraparticular diffusion model.

REFERENCES

1. Sheng, P.X.; Ting, Y.P.; Chen, J.P.; Hong, L. (2004), "Sorption of lead, copper, cadmium, zinc and nickel by marine algal biomass: characterization of biosorptive capacity and investigation of mechanisms", *J. Colloid Interface Sci. 27*, pp. 131-141.
2. Y.S. Yun, D. Park, J.M. park, B. Volesky (2001), "Biosorption of trivalent chromium on the brown seaweed biomass", *Environ. Sci. Technol. 35*, pp. 4353-4358.

3. K.J. Tiemann, J.L. Gardea-Torresdey, G. Gamez, K. Dokken, S. Sias, (1999), "Use of X-ray absorption spectroscopy and esterification to investigate Cr (III) and Ni (II) ligands in alfalfa biomass", *Environ. Sci. Technol. 33*, pp. 150-154.
4. Matlock, M.M., Howerton, B.S., Atwood, D.A. (2002), "Chemical precipitation of heavy metals from acid mine drainage", *Water Res. 36*, pp. 4757-4764.
5. Schlichter, B., Mavrov, V., Erwe, T., Chmiel, H., (1998), "Regeneration of bonding agents loaded with heavy metals by electrodialysis with bipolar membranes", *J. Membr. Sci 232,* pp. 99-105.
6. Fillipi, B.R., Scamehorn, J.F., Christian, S.D., Taylor, R.W. (1998), "A comparative economic analysis of copper removal from water by ligand modified micellar-enhanced ultrafiltration and by conventional solvent extraction", *J. Membr. Sci. 145*, pp. 7-44.
7. Cañizares, P., Pérez, Á., Camarillo, R. (2002), "Recovery of heavy metals by means of ultrafiltration with water soluble polymers: calculation of design parameters", *Desalination 144*, pp. 279-285.
8. Valverde, J.L., de Lucas, A., González, M., Rodríguez, J.F. (2001), "Ion-Exchange equilibria of Cu^{2+},Cd^{2+},Zn^{+2} and Na^{+1} ions on the cationic exchanger Amberlite IR-120", *J. Chem. Eng. Data 46*, pp. 1404-1409.
9. Martins, R.J.E., Pardo, R., Boaventura, R.A.R. (2004), "Cadmium (II) and Zinc (II) adsorption by the aquatic moss fontinalis antipyretica: effect of temperature, pH and water hardness", *Water Res. 38*, pp. 693-699.
10. Cañizares-Vilanueva, R.O. (2000), "Biosorción de metales pesados mediante el uso de biomasa microbiana", *Rev. Lat. de Microbiol. 42*, pp. 131-143.
11. Lin, Z., Wu, J., Xue, R., Yang, Y. (2005), "Characterisation of Au^{3+} biosorption by waste biomass of Saccharomyces cerevisiae", *Spectrochim. Acta A61*, pp. 761-765.
12. Pagnanelli, F., Esposito, A., Veglio, F. (2002), "Multi-metallic modeling for biosorption of binary systems", *Water Res. 36*, pp. 4095-4105.
13. García Pavón, S., "Mitigación del impacto ambiental que generan los residuales sólidos del beneficio de café a partir de la producción de abono orgánico". http://www.cubasolar.cu/biblioteca/Ecosolar/Ecosolar09/HTML/articulo05.htm
14. Rodríguez, N., Zuluaga, J. (1994), "Cultivo de *Pleurotus pulmonarius* en pulpa de café", *Rev. Cenicafé 45*, pp. 81-92.
15. Dávila, M.T., Ramírez, C.A. (1996), "Lombricultura en pulpa de café", *Avances Técnicos Cenicafé 225*, pp. 1-12.
16. Moya-Portuguéz, M., Durán, M., Sibaja, M. (1990), "Obtención de derivados celulósicos a partir de desechos de café", *Agron. Costarrricense 14,* pp. 169-174.
17. Calle, V. H. (1977), "Subproductos del café", *Cenicafé (Boletín Técnico N° 6)*, pp. 84.
18. Tobin, J.M., Cooper, D.G., Neufeld, R.J. (1984), "Uptake of metal ions by Rhizopus arrhizus biomass", *Appl. Environ. Microbiol. 47,* pp. 821-824.
19. Niu, H., Xu, X.S., Wang, J.H. (1993), *"Removal of lead from aqueous solutions by penicillium biomass", Biotechnol Bioeng. 42*, pp. 785–787.
20. DeMarco, M.J., SenGupta, A.K., Greenleaf, J.E. (2003), "Arsenic removal using a polymeric/inorganic hybrid sorbent", *Water Res. 37*, pp. 164-176.
21. Holan, Z.R., Volesky, B. (1995), "Accumulation of cadmium, lead and nickel by fungal and wood biosorbents", *Appl. Biochem. Biotechnol. 53,* pp. 135-145.
22. Langmuir, I. (1918), "The adsorption of gases on plane surfaces of glass, mica and platinum", *J. Am. Chem. Soc. 40*, pp. 1361.
23. Freundlich, H. (1907), "Ueber die adsorption in loesungen", Z. *Phys. Chem*. 57, pp. 385-470.

24. Utomo, H.D., Hunter, K.A. (2006), "Adsorption of heavy metals by exhausted coffee grounds as a potential treatment method for waste waters", *e-J. Surf. Sci. Nanotech. 4*, pp. 504-506.
25. Allen, J.S., Gan, Q., Matthews, R., Johnson, A. (2005), "Mass transfer processes in the adsorption of basic dyes by peanut hulls", *Ind. Eng. Chem. Res.* 44, pp. 1942-1949.
26. Jansson-Charrier, M., Guibal, E., Roussy, J., Delanghe, B., Lecloirec, P. (1996), "Vanadium (IV) sorption by chitosan: Kinetics and equilibrium", *Water Res. 30*, pp. 465-475.
27. Schenker, S., Handschin, S., Frey, B., Perren, R., Escher, F. (2000), "Pore structure of coffee beans affected by roasting conditions", *J. Food Sci. 65*, pp. 452-457.
28. Singh, T.S., Pant, K.K. (2006), "Kinetic and mass transfer studies on the adsorption of arsenic onto activated alumina and iron oxide impregnated activated alumina", *Water Qual. Res. J. Canada* 41, pp. 147-156.
29. Baral, S.S., Das, S.N., Chaudhury, G.R., Rath, P., (2008), "Adsorption of Cr (VI) by treated weed Salvinia cucullata: kinetics and mechanism". Ads. J. the Int. Ads. Soc. 14(1), 111-121.
30. Crank, J., 1975. The mathematics of diffusion. Clarendon Press: Oxford
31. Lagergren, S. (1898), "Zur theorie der sogenannten adsorption gel oster stoffe". *Vetenskapsakad. Handl. 24*, pp. 1-39.

(594c) Development of Waste Water Treatment Process with Sorbents Derived From Waste Concrete Sludge

Akihiro Yamasaki[1], Takeshi Sasaki[1], Yusuke Tsunashima[1], Atsushi Iizuka[2], Yuka Sakai[1], Teruhisa Hongo[3] and Yukio Yanagisawa[4], (1)Department of Materials and Life Science, Seikei University, Musashino, Tokyo, Japan, (2)Research Center for Sustainable Materials & Engineering, Institute of Multidisciplinary Research for Advanced Materials, Tohoku University, Sendai, Miyagi, Japan, (3) Seikei University, Musashino, Tokyo, Japan, (4)The University of Tokyo, Tokyo, Japan

Sorbents for ion removal in waste water derived from waste concrete were developed and their ion removal performances were investigated. The waste concrete sludge is a slurry mixture of cement, aggregate and an industrial waste generated from concrete-using industry such as building. Because waste concrete sludge is rich in calcium components, it can be used as a low-cost material for preparing sorbents potentially applicable to waste water treatment processes

After adding or removal some components in the concrete sludge, the slurry was dehydrate and dried to form solid sorbents. The ion removal performances of the sorbents were examined with simulated waste water. A sorbent with partially extraction of calcium component showed an excellent capture performance for phosphorous ions in water. This result indicates that the sorbent can be applied for the recovery of phosphorus from waste water effluents. The obtained product is composed of hydroxyapatite (HAP), and it showed an excellent sorption performance with heavy metal ions such as lead, cadmium. The sorbent itself also showed an excellent removal performance for boric ions: in this case boric ions can be captured with the formation of ettringite crystal in the sorbent upon mixing with water. The ettringite-containing sorbent can be prepared from the waste concrete sludge by adding aluminum sulfate. The obtained sorbent showed a high removal performance for anions such as boric ions and fluoride ions. Furthermore, the calcium content could be used for fixation of carbon dioxide in the form of calcoum carbonate: an altanate process of carbon sequestration.

See more of this Session: Low Cost Adsorbents for Water Purification
See more of this Group/Topical: Topical 1: Water Technology for Developed and Developing Countries (see also Separations Division)

Adsorption of phenol from aqueous solutions using activated carbon prepared from crofton weed

Chao Hua[1], Runhu Zhang[2], Bai Fang[1],

(1. Key Laboratory of Green Process and Engineering, .Institute of Process Engineering, Chinese Academy of Sciences, Beijing 100190, China

2.Department of Chemical Engineering ,Kunming Metallurgy College ,Kunming 650033 ,China)

Abstract :Activated carbon from crofton weed (ACCW) was evaluated for its ability to remove phenol from aqueous solutions. Batch experiments were used to examine kinetics, adsorption isotherm, pH effect, and thermodynamic parameters. Adsorption data for phenol uptake by ACCW were analyzed according to Langmuir, Freundlich and Temkin adsorption models. Thermodynamic parameters for the adsorption system were determinated at 293 K, 313 K and 333 K ($\Delta H^{\circ} = -22.92$ kJ mol^{-1}; $\Delta G^{O} = -2.87$ to -5.28 kJ mol^{-1} and $\Delta S^{O} = -60.20$ J K^{-1} mol^{-1}). ΔG° values obtained were negative, indicating that the adsorption of the phenol on the surface of ACCW was a spontaneous adsorption process. The kinetics of this process was described very well by a pseudo-second-order rate equation. These results show that the ACCW could be considered as a potential adsorbent for phenol in aqueous solutions.

Keywords: Phenol; Adsorption; Crofton weed; Isotherms; Activated carbon

(594e) Metals Adsorption From Bioleaching Solution of Printed Circuit Board

Len Foong Koong, Gordon Mckay, John Barford and Koon Fung Lam, Chemical and Biomolecular Engineering, The Hong Kong University of Science and Technology, Clear Water Bay, Hong Kong

Adsorption of Metals from Bioleaching Solution of Printed Circuit Board

Len Foong KOONG, Gordon MCKAY, John BARFORD

The Hong Kong University of Science and Technology, Clear Water Bay, Hong Kong

Printed Circuit Boards (PCB) are used to physically support and electrically connect electronic components via copper conductive pathways. They are composed of polymers, glass fiber, solder and metals, most of which are hazardous to the environment if not disposed or treated carefully. In the last two decades electrical and electronic products have become more affordable and better, due to the rapid innovation of the industry. As a result, the product life becomes shorter and more electronic waste is generated, adding stress to the environment, landfill space and public health. Thus it has become of great interest to improve the recycling of waste PCBs. The current technology first removes recoverable components via manual or robotic disassembly line; the materials on the scrapped PCBs are then recycled by mechanical separation, pyrometallurgy or hydrometallurgy. Bioleaching of PCBs is gaining more attention recently as a more environmental friendly and cheaper treatment for PCBs. Researchers have found that the mixture of bacteria Acidithiobacillus Ferrooxidans (A. ferrooxidans) and Acidithiobacillus Thiooxidans (A. thiooxidans) can effectively leach out metals from PCBs, notably copper, nickel, zinc, lead and tin. As the culture mixture produces sulfuric acid, lead and tin will precipitate as sulfate salts. However there is still a need to recover the remaining metals from the bioleaching solution. Two types of adsorbents: synthetic compound NH2-MCM-41 and commercially available Chitosan are used in the single and binary components isotherm studies. NH2-MCM-41, a type of mesoporous silica grafted with aminopropyl groups and known for its heterogeneity of surface functional group is used to compare with the selectivity (binary components studies) and capacity of Chitosan (single components studies) which contains primary amine as functional group. Ultimately it is our goal to separate the metals individually to make downstream recovery easier.

See more of this Session: Low Cost Adsorbents for Water Purification
See more of this Group/Topical: Topical 1: Water Technology for Developed and Developing Countries (see also Separations Division)

Ultrafiltration/Nanofiltration Based Technique for the Removal of Endocrine Disruptors from Surface Water

Abstract ID# 222410

Ana Gil, Georgina Sarmento, Miguel Minhalma, Elizabeth Duarte and Maria Norberta De Pinho

Introduction

Membrane technologies and more specifically pressure-driven membrane processes like microfiltration (MF), ultrafiltration (UF), nanofiltration (NF) and reverse osmosis (RO) (1 - 9) are playing a crucial role on the production of drinking water of high quality. This is attributed to the fact that these technologies address in a very specific way: 1) The efficient removal of residual organic matter (NOM) that acts as a precursor of the formation of disinfection by-products that are health hazards; 2) The removal of conventional micropollutants such as pesticides, alkylphthalates and endocrine disruptors chemicals (EDs). The EDs and namely the 4-Nonylphenil (NP) can be integrated in surface waters through diverse mechanisms, like direct discharge of industrial and domestic wastes, the discharge of effluents from biological treatment plants and drains of agriculture into rivers and soils.

The present work addresses both the removal of NOM by Ultrafiltration and the removal of NP by Nanofiltration. Two sets of experiments were carried out in concentration mode: i) Ultrafiltration of surface water from Tagus River (Valadas, Portugal); ii) Nanofiltration of surface water from Sado River (Setúbal, Portugal).

Results

Ultrafiltration of Tagus River Water (10)

The variation of the permeation fluxes with the water recovery rates is displayed in Fig. 1.

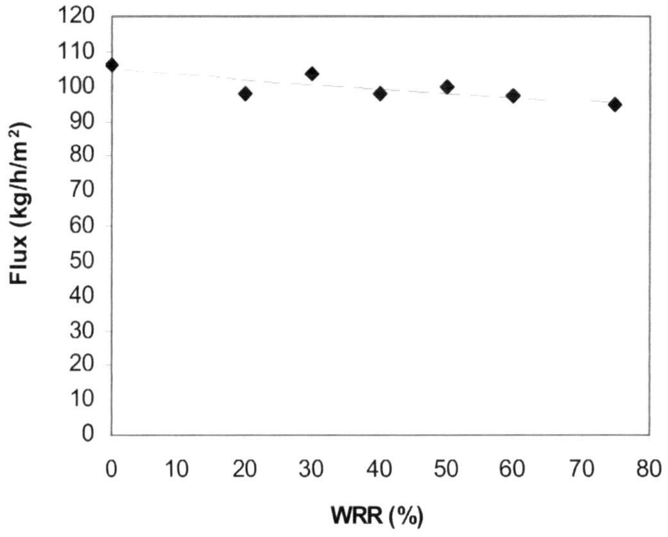

Figure 1. Permeation fluxes of Tagus River water as a function of water recovery rates. Membrane: ETNA10PP; Operating Conditions: Transmembrane Pressure = 4 bar; Feed flow rate = 554 L/h.

The rejection coefficients to the total organic carbon (TOC) and to the dissolved humic substances (UVA*$_{254nm}$) are displayed in Fig. 2 in a wide range of water recovery rates.

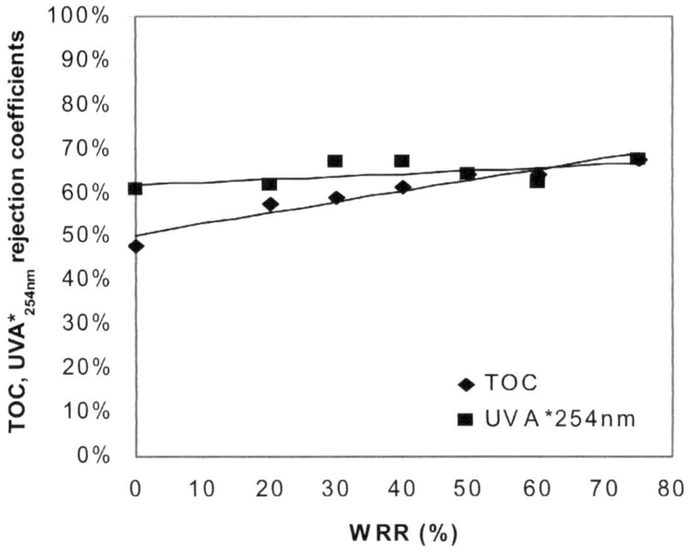

Figure 2. Organic matter rejection coefficients as a function of water recovery rates.

Membrane: ETNA10PP; Operating conditions: Transmembrane Pressure = 4 bar; Feed flow rate = 554 L/h.

Nanofiltration of Sado River Water (11)

The decline of the permeation fluxes with the increase of the water recovery rates is displayed in Figures 3 and 4 for surface water without and with inoculation of 4-Nonylphenol, respectively.

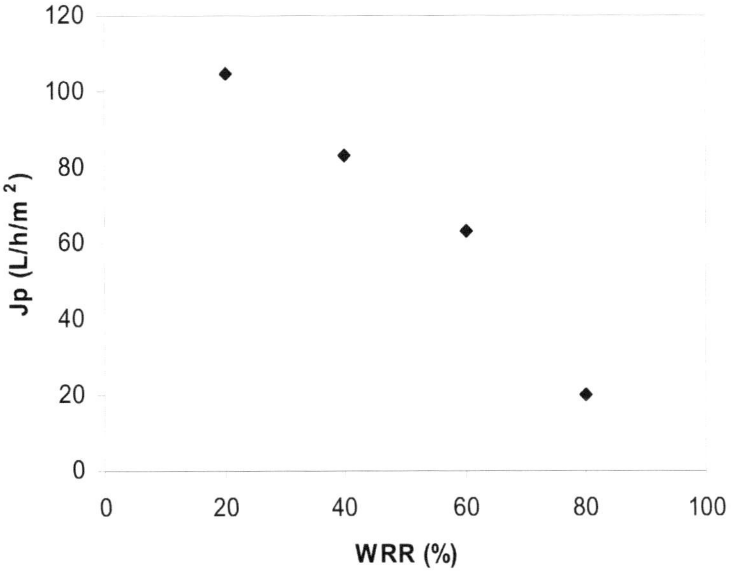

Figure 3. NF permeate fluxes as a function of water recovery rates for the Raw Sado River water. Membrane: NF 270; Transmembrane pressure = 30 bar.

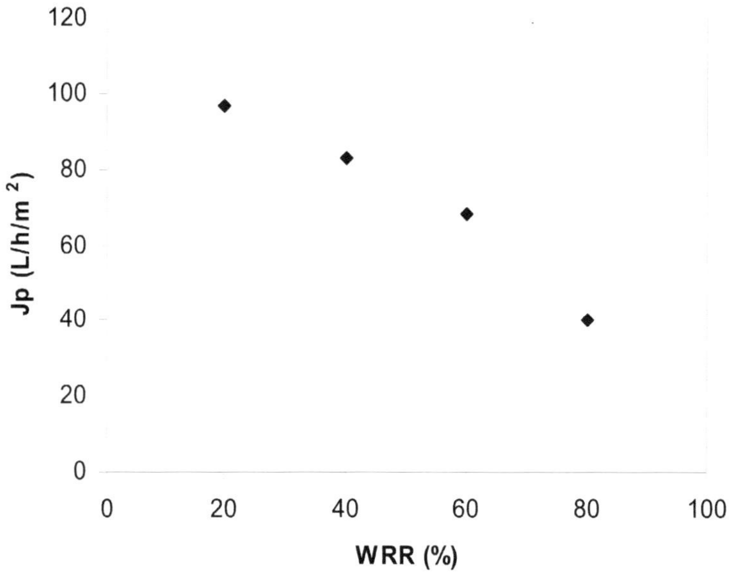

Figure 4. NF permeate fluxes as a function of water recovery rates for the Sado River water inoculated with 4-Nonylphenol.

Membrane: NF 270; Transmembrane pressure = 30 bar.

In the nanofiltration of Sado River water with and without inoculation of NP the rejection coefficients to NOM are 77 % and to NP are 100 %.

Conclusions

The decline of the surface water permeation fluxes with the water recovery rates is more pronounced in the operation of nanofiltration than in the operation of ultrafiltration.

In nanofiltration the flux decline is less pronounced when the water is inoculated with NP. At the water recovery rate of 80 % the permeate flux is 40 $L/h/m^2$ for water inoculated with NP and 20 $L/h/m^2$ for the raw water.

The average apparent rejection coefficient obtained for NOM is 77 % and for the endocrine disruptor, NP, is 100%, for the surface water with and without inoculation of the endocrine disruptor. The higher rejection coefficients to endocrine disruptors when compared to the ones of natural organic matter give evidence that there is a NOM fraction of high molecular weight retaining the endocrine disruptor in the concentrate stream and a permeate stream with NOM of lower molecular weight that is free of endocrine disruptor.

References

(1) Hong, S. and Elimelech, M., Chemical and physical aspects of natural organic matter (NOM) fouling of nanofiltration membranes, J. Membr. Sci., 132 (1997) 159-181.

(2) Cho, J.W., Amy, G. and Pellegrino, J., Membrane filtration of natural organic matter: initial comparison of rejection and flux decline characteristics with ultrafiltration and nanofiltration membranes, Water Res., 33 (11) (1999) 2517-2526.

(3) Kiso,Y., Sugiura, Y., Kitao, T. and Nishimura, K., Effects of hydrophobicity and molecular size on rejection of aromatic pesticides with nanofiltration membranes, J. Membr. Sci., 192 (1-2) (2001) 1-10.

(4) Kiso, Y., Mizuno, A., Othman, R., Jung, Y.J., Kumano, A. and Ariji, A., Rejection properties of pesticides with a hollow fiber NF membrane (HNF-1), Desalination, 143 (2) (2002) 147-157.

(5) Costa, A.R., de Pinho, M.N., The role of membrane morphology on ultrafiltration for natural organic matter removal, Desalination, 145 (1-3) (2002) 299 - 304.

(6) Costa, A.R., de Pinho, M.N., Coagulation/flocculation/ultrafiltration for natural organic matter removal in drinking water production, Water Science and Technology: Water Supply, 4-5 (2004) 215-222.

(7) Costa, A.R., de Pinho, M.N., Performance and cost estimation of nanofiltration for surface water treatment in drinking water production, Desalination, 196 (1-3) (2006) 55-65.

(8) Costa, A.R., de Pinho, M.N., Elimelech, M., Mechanisms of colloidal natural organic matter fouling in ultrafiltration, J. Membr. Sci., 281 (2006) 716–725.

(9) Jung, Y.J., Kiso, Y., Park, H.J., Nishioka, K. and Min, K.S., Rejection properties of NF membranes for alkylphenols, Desalination, 202 (1-3) (2007) 278-285.

(10) Costa, A. R., Mechanisms of membrane fouling by colloidal organic matter and optimization of membrane processes for surface water treatment, PhD Thesis (2007), Instituto Superior Técnico, Technical University of Lisbon, Portugal.

(11) Salvaterra, A. F., Sarmento, G., Minhalma, M., and De Pinho, M. N., Nanofiltration of surface water for the removal of endocrine disruptors, Desalination and Water Treatment (2011), In press.

(658b) The Importance of a Well-Designed Cross-Flow System In the Removal Study of Adsorbing Estrogens by Nanofiltration Membranes

Andrea JC Semiao, School of Engineering, The University of Edinburgh, Edinburgh, United Kingdom

Life on Earth, including the survival of humans and other living species, is strongly dependent on the production of clean water for potable usage. Endocrine disruptors, such as hormones, have a potential harmful effect on human health and should therefore be removed from potable water resources. However, so far, such disruptors have not been completely removed by conventional wastewater treatment processes and are consequently discharged into potable water resources, contaminating them. On the other hand, nanofiltration (NF) is widely used in water treatment as is the case of the Mery-sur-Oise plant in Paris that treats the water from the Seine River to recover it as potable water, through the removal of pesticides and other contaminants. Both the difficulty in removing the endocrine disruptors that exhibit a potential health risk and the broad use of NF for waste water treatment make the understanding of the removal mechanisms of the above-mentioned pollutants a crucial issue.

At laboratory scale, cross-flow systems with a slit channel have been widely used in the research of membrane water treatment, particularly when the focus lies on removal of trace contaminants. Slits constitute a laboratory model of spiral-wound modules allowing for the study of the fundamentals of mass and momentum transport mechanisms in the feed channel by controlling its hydrodynamics. Spiral-wound modules are, in turn, the most common membrane geometry used in water treatment, due to their compact geometry with high surface/volume ratio.

This work aims at demonstrating the importance of having a well-designed cross-flow system and without oil leakages from the pumping system in the removal of endocrine disruptors by showing how both an ill-designed cross-flow system and the presence of oil in the aqueous solution impact on the results of their removal by NF membranes. For that, a comparison between hormone filtration results yielded by both type of systems is performed. Additionally, further experiments performed with the well-designed system aim at defining, understanding and quantifying the physical mechanisms of momentum and mass transfer controlling the adsorption of hormones onto polymeric NF membranes, viewing the development of a predictive model.

In an ill-designed system, pump delivering flow rates to the cross-flow membrane cell above a limit value (Reynolds number based on the hydraulic slit diameter, Re_{dh}, above 10,000) causes the membrane active layer to be ripped off from the support layer, and yields a E2 retention below 15% whereas that obtained for a Re_{dh} number of 2800 in a well-designed cross-flow system is higher than 80%. On the other hand, the presence in the aqueous solution of oil contamination originated from the pump of the ill-

designed system caused the E2 retention to be much lower than expected: 25% in the presence of oil against 65% in the absence of oil obtained in the well-designed system. These and other results led to the abandonment of the ill-designed system since, in opposition to it, the well-designed one allowed to mimic spiral-wound hydrodynamic conditions and carry out the work aiming at understanding the fundamentals of the removal mechanisms of hormones by NF membranes. Special care was taken in the new system to re-design the pump head of a commonly used diaphragm pump in cross-flow systems to make sure that there was no oil leakage. These findings are very important for pump manufacturers and users.

Once these changes were made, it was possible to proceed with the experiments viewing the definition of a predictive model of adsorption of hormones onto polymeric NF membranes. First, knowledge from experiments was acquired on how hydrodynamics and operating conditions - pressure, circulating Reynolds numbers (Re_h) and feed concentration - affect hormone adsorption: adsorption is governed by the initial concentration at the membrane surface that, in turn, depends on the hormone feed concentration, operating Re_h and pressure. Membrane retention, however, depends on the initial polarisation modulus, given by the fraction of the initial concentration at the membrane surface divided by the initial feed concentration. Then, based on these results, a sorption model was developed, predicting very well both feed and permeate transient concentrations for two different hormones in the normal range of operating pressures and Re_h of spiral-wound membrane modules. These findings are an important advancement in determining which membrane would be more suitable to effectively remove hormones since the model predicts well the feed and permeate concentrations with a substantial reduction of experimental work.

See more of this Session: Micropollutants, Disinfection, and Other Considerations in Water Treatment
See more of this Group/Topical: Topical 1: Water Technology for Developed and Developing Countries (see also Separations Division)

(658c) SW-RO Membranes for the Removal of Ammonium Thiocyanate From Coking Wastewaters In Steel Manufacturing

Susana Luque[1], **Jose R. lvarez**[1], M. El-Mansour[1], F.E. Antn[1], B. Gonzlez[2] and B. P. Vivas[2], (1) Department of Chemical and Environmental Engineering, University of Oviedo, Oviedo, Spain, (2) ArcelorMittal Research & Development, Global R&D Asturias, Avils, Spain

The liquid effluents from a coke plant in an integrated steel manufacturing facility usually contain ammonium cation, thiocyanate and cyanide anions, phenols and suspended solids. Cyanides, cyanates and thiocyanates are hazardous substances with a negative effect on living species and the environment. The generation of these pollutants is inherent to the manufacturing process, and the wastewaters containing them cannot be discharged without treatment, according to the current regulations. Chemical treatment, incineration and biological digestion are the most common treatment methods. However, when the thiocyanate ion concentration exceeds 400 ppm, the microbial activity is hindered, and thus, an additional chemical pre-treatment is needed. Incineration can also be used to treat concentrated thiocyanate effluents, but at a high cost.

Reverse osmosis (RO) has proved to be a feasible technology in the recovery of thiocyanate from the aforementioned condensates. Several experiments have been carried out at laboratory scale to choose a suitable membrane and to gather experimental data which would allow developing a model to design and scale up the filtration process. Commercially available RO spiral wound membranes were selected. The membrane showing the best behaviour was the so-called SW30-HR membrane from DOW FILMTECTM as it gave adequate values of rejection and permeability. The effect of process parameters has also been addressed and recommendations for plant operation have been proposed.

Onsite pilot plant runs have also been conducted in order to evaluate RO performance when operating longer term with the actual industrial stream. A pilot unit equipped with two 4" spiral wound modules were used. The results have been in close agreement with those obtained at laboratory scale.

A mathematical model which represents the process has then been proposed, validated and used to design a full scale plant which accomplishes the desired separation. Coking wastewaters effluents usually have thiocyanate concentrations between 1000 and 3000 ppm, but eventually, it is possible to have condensates that reach 9000 ppm. Therefore, several case studies and sensitivity analysis have been carried out based on both these typical and extreme values. An economic evaluation is also included.

ACKNOWLEDGMENTS

Financial support by PCTI (RIA Project, PEST08, Principality of Asturias, Spain)

(658d) Environmental and Economic Evaluation of Wastewater and Water Disinfection Technologies

Tapas K. Das, School of Engineering, Saint Martin's University, Lacey, WA

This paper describes some of the common use of disinfectants and disinfection practices, along with most modern and emerging technologies that are proving to be more efficient in disinfecting varieties of microorganisms present in wastewater and drinking water. In rural areas of developing nations waterborne disease such as typhoid, cholera, hepatitis, and gastroenteritis, infect and kill many infants and children each day, according to the World Health Organization (WHO). With increasing emphasis on promoting a sustainable water ecological future and concern over introducing a toxic chemical in the water, the design of the disinfection process is increasingly leaning toward technologies that destroy the pathogens while balancing the effect of this disinfected wastewater on the population of aquatic biota or a drinking water supply. Disinfection is the most important water treatment process for preventing the spread of infectious diseases. For the past several decades, chlorine has been the disinfectant of choice. The discovery and subsequent regulation of some chemical by-products of chlorine (e.g., trihalomethanes) has increased the popularity of other disinfectants such as ozone, hydrogen peroxide, UV radiation, sonication, etc. In this presentation, UV radiation will be used as the primary disinfection method, and will be tested in conjunction with chlorine , chlorine dioxide, sodium hypochlorite Fenton reagent and O_3 to enhance the germicidal effectiveness of protozoa, *cryptosporidium parvum*, *giardia duodenalis cysts*, and other microorganisms. The effectiveness of removing pathogens, odorous compounds/color and the environmental and economic impacts of wastewater and drinking water disinfection technologies will be assessed.

See more of this Session: Micropollutants, Disinfection, and Other Considerations in Water Treatment
See more of this Group/Topical: Topical 1: Water Technology for Developed and Developing Countries (see also Separations Division)

(658e) Chemical Properties and Degradation Efficiency of Radio-Frequency Plasmas In Water

Selma Mededovic, Dept. of Chemical and Biomolecular Engineering, Clarkson University, Potsdam, NY

To treat contaminated water, advanced oxidationtechnologies (AOT) utilize various precursors and processes to produce hydroxylradicals (OH); species which are, because of their highoxidation potential, capable of degrading virtually any organic compound in theaqueous solution. However, in some of the AOTs, the production of OHcanbe quite limited depending on, for example, the typeof the OHprecursor (H_2O vs. H_2O_2)and the turbidity of the solution. Thus, there is a requirement for atechnology that will overcome the drawbacks associated with the current AOTsand the preliminary results from this study indicate that radio-frequency (RF)plasma in water, a new "type" of plasma formed by low-voltage high-frequency(13.56 MHz) electrical discharge, could be that technology. Compared to theconventional high-voltage streamer-like discharges, the volume of the RF plasmais several times larger and the intensity of the plasma-generated UV light appearsto be higher. Thus, the goal of this study is to chemically and spectroscopicallycharacterize RF plasma and asses its efficiency for the degradation ofRhodamine B dye.

See more of this Session: Micropollutants, Disinfection, and Other Considerations in Water Treatment
See more of this Group/Topical: Topical 1: Water Technology for Developed and Developing Countries (see also Separations Division)

Determination of Bisphenols Using Ultraviolet Spectroscopy

Alfred R. Conklin, Jessica Veite, Jennifer Wilson

Chemistry Department, Wilmington College Wilmington, Ohio

Background

Bisphenols are an environmental concern because of both their wide use and evidence connecting them to biological effects especially relating to their estrogenic activity (1, 2). There are a family of bisphenols which are commonly used as monomers in common plastics, including drink containers and liners for drink and food cans. Bisphenols can leach out of plastic containers into their contents. When consumed the consumers are exposed to bisphenols.

Exposure to bisphenols has been related to a change in the size of reproductive organs, sperm production, and the onset of puberty. This has lead to concern over exposure to bisphenols particularly in young children (3,4,5).

The most commonly cited bisphenol is Bisphenol A, see Figure 1. There is however a family of bisphenols derived by making substitution on the methylene group connecting the two phenol moieties. As is seen in the Figure, bisphenols contain two phenol moieties.

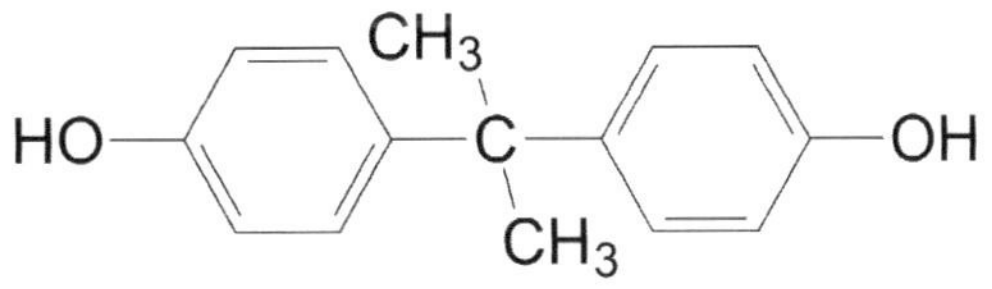

Figure 1. Structure of Bisphenol A.

Gas chromatography-mass spectroscopy (GC-MS) has been used to determine the bisphenol content of foods, including water (6). Solutions containing bisphenol can be injected directly into the GC-MS and analyzed. If the concentration is low BPA can be absorbed on an appropriate absorbent and concentrated before analysis (7, 8). It is also possible to dervatize the bisphenol before analysis (9). High performance liquid chromatography HPLC has been used to isolate and detect bishenols. (10,11). Molecularly imprinted polymers along with GC and HPLC and stable isotopes have also been used in bisphenol analysis (12, 13, 14).

Because phenol has an ultraviolet spectrum it is to be expected that bisphenol would also have an ultraviolet spectrum and that the absorptivity would be high. This is indeed the case. It is thus possible to determine the amount of bisphenols in both aqueous and ethanolic solutions using ultraviolet spectroscopy.

In many cases plastics are used in the bottled water industry so using ultraviolet spectroscpy to determine the bisphenols in aqueous solutions should be possible. This type of analysis is simple and requires few resources. If additional confirmation of the existence or quantity of bisphenol is needed a GC-MS analysis can be carried out.

Experimental.

Bisphenol A [2,2-Bis(4-hydroxyphenyl)ethane] 97% was obtained from Sigma Aldrich as small spheres. Spheres were ground to a fine powder, using a mortar and pestle before use.

Deionized Distilled water was prepared and stored in glass until used (DID water).

Absolute Ethanol 90.25 % containing 4.75% methanol and 5.00% isopropanol as denaturants was used in all experiments involving ethanol.

Syringe filters, Nylon, cellulose acetate, PVDF, PTFE, PSE with 0.45 and 0.2 μm, pores were obtained from Fisher Scientific.

Ultraviolet spectroscopy was carried out using Quartz, 10mm path length, 0.7 ml cells. Ultraviolet spectra were obtained using a Shimadzu UV-1650PC spectrophotometer. GC-MS analysis was carried using a Thermo ISQ with a Focus GC and a TG-5MS 30m x 0.25mm x 0.25μm capillary column. The GC inlet temperature was set at 220 °C and the transfer line at 250 °C with a helium carrier gas flow rate of 1.2 ml per minute. The oven was set at 125 °C for 0.5 minutes then increased to 250 °C at 20 °C/minutes and held at this temperature for 2 minutes. 1μl injections were used for all samples.

Procedure

Solutions containing varying amounts of BPA but typically in the range of 2 to 200 ppm bisphenol A in both DID and ethanol were prepared. Aqueous solutions were analyzed using ultraviolet spectroscopy. Ethanolic solutions were analyzed using both ultraviolet spectroscopy and GC-MS. In both cases standard curves were prepared using Excel and the linear fit for the lines and their r^2 calculated.

BPA standards were filtered through various syringe filters and the ultraviolet absorption before and after filtration measured. Each standard was filtered three different times, except as noted, and the standard deviation determined.

Results

Both aqueous and ethanolic BPA solutions analyzed by ultraviolet spectroscopy and GC-MS produced standard curves with $r^2 = 0.99$, see Figure 2. Small additions of either water to ethanolic, or ethanol to aqueous solution resulted in small but significant changes in absorption.

Absorption of BPA from the standard solution by syringe filters varied between 96% to 4 % of the standard BPA solution contained a calculated 175 mg BPA per liter solution. This

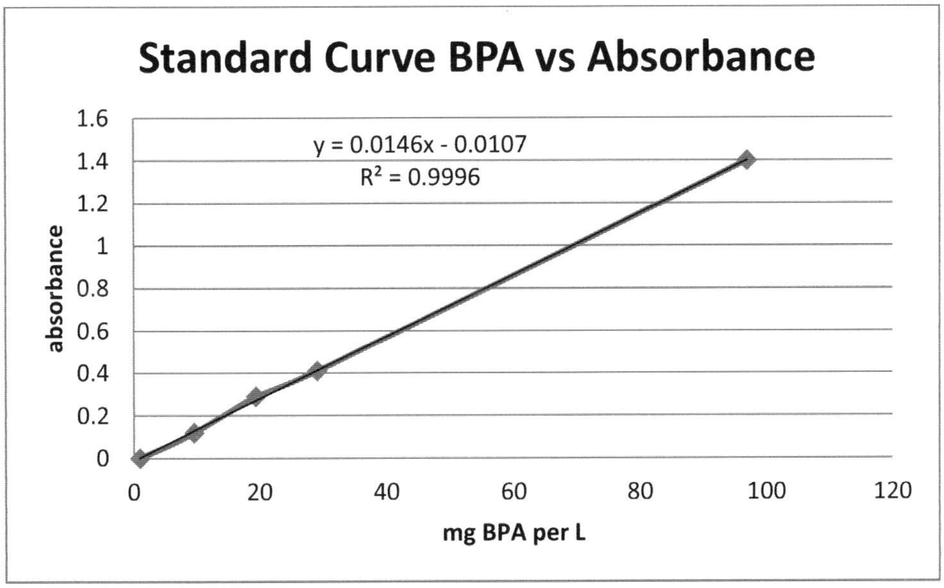

Figure 1. An example of a graph of BPA vs absorption at 277 nm. Note that although the abosrpance for 100 mg/l is outside the normally used range of ultraviolet spectroscopy the r^2 is still 0.999.

Table 1. Adsorbance of BPA by various syringe filters containing different filter material

Filter Brand	Material	Pore Size	BPA Adsorbance after filtration	Standard Deviation
None Standard BPA Solution(1)*			2.51 (2)*	±0.03
Whatman	Nylon	0.45	1.51	±0.2
Fisher	Nylon	0.45	0.11	±0.06
Gelman	Nylon	0.45	0.78	±0.3 (3)*
Whatman	Cellulose Acetate	0.2	0.64	±0.02
Whatman	Cellulose Acetate with GMF	0.2	0.26	±0.1
Whatman	PVDF	0.2	1.51	±0.3
Whatman	PTFE	0.2	2.11	±0.1
Millex	PVDF	0.22	0.74	±0.1
Millex	PTFE	0.45	2.44	±0.06
Corning	PSE	0.2	0.94	±0.7(3)*

Notes: (1)* an aqueous solution
(2)* standard solution was not filtered.
(3)* initial measurements showed large variation so 6 replicates were determined.

solution was chosen so that the syringe filters absorbing the most BPA would still provide results which would be on scale and could be graphed. Solutions filtered through Fisher Nylon had the lowest adsorbance (lowest amount of BPA) after filtration, and Millex PTEF had the highest absorption after filtration (highest amount of BPA) but was not significantly different from the control. Filtering with Whitman PTEF and Nylon also produced a filtrate with high adsorbance. All other results were significantly lower (statistically) than the control although not always significantly different from each other. Variations in pore size did not seem to affect adsorbance although source of filter material did. Table 1 gives complete data for various filters tested. All measurements were replicated 3 times except as noted.

Conclusion

Ultraviolet spectroscopy can be used to determine the occurrence and concentration of Bisphenols in simple solutions. Caution must be used in using ultraviolet spectroscopy when the extractant may be contaminated with another solvent or extractant. Also it is possible that some components of both drinks and foods may interfere with either the λ_{max} or absorption. However with relativly pure water samples, ultraviolet spectroscopy works well. In cases where there is some doubt GC-MS can be used to confirm the results.

Syringe filters adsorb Bisphenol, with PTFE generally adsorbing the least and Nylon adsorbing the most. Syringe filter filtration should be avoided when determining the concentration of BPA in solutions. If syringe filtration must be done the results must be corrected for the adsorption of BPA by the filters.

References

1. Vom Saal, Frederick S., Paul S. Cooke, David L. Buchanan, Paola Palanza, Kristina A. Thayer, Susan C. Nagel, Stefano Parmigiani, Wade V. Welshons (1998), "A Physiologically Based Approach To the Study of Bisphenol a and Other Estrogenic Chemicals On the Size of Reproductive Organs, Daily Sperm Production, and Behavior," Toxicology and Industrial Health, 14:239-260.

2. Knaak James B. and Lloyd J. Sullivan (1966), "Metabolism of bisphenol A in the rat," Toxicology and Applied Pharmacology 8:175-184.

3. Howdeshell, Kembra L, Andrew K. Hotchkiss, Kristina A. Thayer, John G. Vandenbergh and Frederick S. vom Saal (1999), "Environmental toxins: Exposure to bisphenol A advances puberty," Nature 401:763-764.

4. Laws, Susan C., Stephan A. Carey, Janet M. Ferrell, Gerald J. Bodman, Ralph L. Cooper (2000), "Estrogenic Activity of Octylphenol, Nonylphenol, Bisphenol A and Methoxychlor in Rats," Toxicol. Sci. 54:154-167.

5. Staples Charles A., Philip B. Dome, Gary M. Klecka, Sondra T. Oblock and Lynne R. Harris (1998), "A review of the environmental fate, effects, and exposures of bisphenol A", Chemosphere 36:2149-2173.

6. Basheer Chanbasha and Hian Kee Lee (2004), "Analysis of endocrine disrupting alkylphenols, chlorophenols and bisphenol-A using hollow fiber-protected liquid-phase microextraction coupled with injection port-derivatization gas chromatography–mass spectrometry," J. Chromatography A, 1057:163-169.

7. Braun P., M. Moeder, St. Schrader, P. Popp, P. Kuschk and W. Engewald (2003), "Trace analysis of technical nonylphenol, bisphenol A and 17α-ethinylestradiol in wastewater using solid-phase microextraction and gas chromatography–mass spectrometry," J. Chromatography A, 988:41-51.

8. Chang Chia-Min, Chi-Chi Chou and Maw-Rong Lee (2005), "Determining leaching of bisphenol A from plastic containers by solid-phase microextraction and gas chromatography–mass spectrometry," Analytica Chimica Acta, 539:41-47.

9. Salafranca Jesús, Ramón Batlle and Cristina Nerín (1999), "Use of solid-phase microextraction for the analysis of bisphenol A and bisphenol A diglycidyl ether in food stimulants," J. Chromatography A, 864:137-144.

10. González-Casado A., N. Navas, M. del Olmo, J.L.Vílchez (1998), "Determination of Bisphenol A in Water by Micro Liquid-Liquid Extraction Followed by Silylation and Gas Chromatography-Mass Spectrometry Analysis," J. Chromatographic Science, 36:565-570.

11. Kataoka Hiroyuki, Megumi Ise, Shizuo Narimatsu (2002), "Automated on-line in-tube solid-phase microextraction coupled with high performance liquid chromatography for the analysis of bisphenol A, alkylphenols, and phthalate esters in foods contacted with plastics," J. of Separation Science, 25:77–85.

12. Szap Peter, Ivar Kesse and Jack Klapp (1978), "The Analysis of Bisphenol a by High Performance Liquid Chromatography," J. Liquid Chromatography 1:89-96.

13. Kawaguchi Migaku, Yoshio Hayatsu, Hisao Nakata, Yumiko Ishii, Rie Ito, Koichi Saito and Hiroyuki Nakazawa (2005), "Molecularly imprinted solid phase extraction using stable isotope labeled compounds as template and liquid chromatography–mass spectrometry for trace analysis of bisphenol A in water sample," Analytica Chimica Acta, 539 83-89.

14. Navarro-Villoslada F., Blanca San, Vicente and María C. Moreno-Bondi (2004), "Application of multivariate analysis to the screening of molecularly imprinted polymers for bisphenol A,"Analytica Chimica Acta, 504:149-162.

(659a) Static Mixing Spacers for Spiral Wound Modules

Ashkan Iranshahi and **Glenn Lipscomb**, Department of Chemical and Environmental Engineering, University of Toledo, Toledo, OH

Membranes play a critical role in many separation processes ranging from nitrogen production to desalination. The membranes are used most commonly in the form of fine hollow fibers or sheets. The sheet form is used to produce spiral wound modules that permit the efficient contacting of process streams with the membrane in a compact form.

Spiral wound modules are the dominant form used in reverse osmosis and nanofiltration membrane processes. Spacers are used in spiral wound modules to create feed and permeate flow channels and to enhance mass transfer rates by reducing concentration polarization. Most spacers possess a ladder or mesh structure and rely on the generation of turbulence and eddies to enhance mass transfer. The literature contains numerous experimental and theoretical studies of how pressure drop and mass transfer rates depend on the geometry of these spacers.

Spacer design features that increase mass transfer rates also tend to increase pressure drop. The preferred spacer is determined by the tradeoff between the reduction in membrane area and concomitant increase in pumping costs. A review of previous spacer design studies is presented with emphasis on the proper basis for comparison of the results.

Following this review, we present a new spacer paradigm. The spacer acts as a static mixer for planar flow channels and mixes without generating turbulence and eddies. Simulations of flow and mass transfer with the new spacer were performed using computational fluid dynamics. The results indicate mass transfer rates are greater than that reported for other spacer designs. Flow visualizations clearly show how the spacer mixes fluid.

See more of this Session: Modeling and Computational Approaches for Water Treatment
See more of this Group/Topical: Topical 1: Water Technology for Developed and Developing Countries (see also Separations Division)

(659b) Atomistic Simulation of Water and Salt Transport In RO Membrane Active Layers

Jed W. Pitera, Young-hye Na, Ankit Vora and Geraud Dubois, Science & Technology, IBM Almaden Research Center, San Jose, CA

As part of a combined computational and experimental effort to develop new RO membrane materials, we have carried out detailed equilibrium and nonequilibrium molecular dynamics simulations of the aromatic polyamide active layers of reverse osmosis desalination membranes. Simulations carried out at low salt concentration yield hydraulic permeabilities in good agreement with experimental results. Statistical issues make direct simulation of salt rejection impractical, but we will describe the observed distribution and transport mechanisms of salt ions in the polymer.

See more of this Session: Modeling and Computational Approaches for Water Treatment
See more of this Group/Topical: Topical 1: Water Technology for Developed and Developing Countries (see also Separations Division)

(659c) Dynamic Data-Driven Models of Reverse Osmosis Desalination Systems

Xavier Pascual[1], Han Gu[2], Alex Bartman[3], Aihua (Richard) Zhu[4], Anditya Rahardianto[2], Jaume Giralt[5], Robert Rallo[6], Joan Ferrer-Gin[6], Francesc Giralt[7] and Yoram Cohen[2], (1)Enginyeria Qumica, Universitat Rovira i Virgili, Tarragona, Spain, (2)Chemical & Biomolecular Engineering, University of California, Los Angeles, Los Angeles, CA, (3)Chemical and Biomolecular Engineering, University of California, Los Angeles, Los Angeles, CA, (4)Chemical and Biomolecular Engineering, UCLA, Los Angeles, CA, (5)Departament d'Enginyeria Qumica, Universitat Rovira i Virgili, Tarragona, Spain, (6) Enginyeria Informatica i Matematiques, Universitat Rovira Virgili, Tarragona, Spain, (7)BIOcenit, Departament d'Enginyeria Qumica, Universitat Rovira i Virgili, Tarragona, Spain

The control and optimization of reverse osmosis (RO) desalination plants can benefit from the use of plant-specific process models that account for feed conditions (salinity, temperature and pH) as well as operational deviation of various plant components (e.g., pumps, valves). In this regard, data-driven models, developed using machine-learning techniques, can be particularly useful for assessing plant performance, operational deterioration (e.g., due to membrane fouling and failure of plant components), as well as forecasting of plant performance based on historical operational data. Accordingly, the present study presents a systematic approach for the development of data-driven models of RO membrane plant operation based on real-time data. The approach was developed based on data acquired from a novel fully-automated and remotely controlled mini-mobile-modular (M3) RO pilot plant designed and built at UCLA. Operational data with the M3 plant were obtained over a wide range of process conditions (e.g., product water recovery, feed salinity, temperature, pH, feed pressure and cross flow velocity). The M3 RO plant rapidly and automatically generated (aided by its embedded computer controlled system), sets of time-series data for prescribed and randomly set cyclical process conditions (including the effect of process perturbations) under both dynamic (trajectory toward control set points) and steady state operations. The acquired plant data (recorded at high frequency) were utilized to develop data-driven models based on Self-Organizing-Maps (SOM) and Support Vector Machines (SVM). Plant sensors (e.g., salinity and flow rate monitors) can fail or malfunction (even if for short periods) and as was also encountered in the present plant. Therefore, various data filters were applied along with assessment of data consistency (based on overall and salt mass balances) for fault detection and isolation (e.g., identification of malfunctioning plant sensors). Subsequently, optimal data-driven plant models were developed using the appropriate feed conditions (as input variables) to predict the transitions between different states of plant operation after changes in the manipulated variables, thereby providing accurate estimates of permeate and retentate conditions (flow-rate and conductivity). An important element in the model development was a variable-length data segment ("window") that was introduced in order to include information of most recent trends of plant operation. The current approach demonstrates that dynamic data-driven RO plant models can be developed at a reasonable level of accuracy, and can be used in tandem with deterministic models to describe the complete domain of plant operability under both steady and unsteady state conditions.

(659d) The Role of External Electric Fields In Enhancing Water Treatment Technologies

Sohail Murad, Chemical Engineering, University of Illinois at Chicago, Chicago, IL

Molecular simulations have been carried out using the method of molecular dynamics to investigate the role of external electric fields on the ion mobility, drift velocity and drift-diffusion rate of ions in aqueous electrolyte solutions. These properties are critical for a range of processes including electro-dialysis, electro-deionization, electro-phoresis and electro-osmosis. Our results show that external electric fields relax the hydrated ion structure at significantly larger time scales (between 300 - 800 ps), than most other relaxation processes in solutions (generally of the order of 1 ps). Previous studies that did not account for the much longer relaxation times did not observe this behavior for ions even with very high electric fields. External electric fields must also overcome several (at least two or more) activation energy barriers to significantly change the structure of hydrated ions. As a result, the dynamic behavior changes almost in bands as a function of electric field strengths, rather than linearly. Finally, the effect of the field is much less dramatic on water than the ions. Thus electric fields will be of more significance in processes that involve the transport of ions (such as electro-deionization), than the transport of water (electro-osmosis)

See more of this Session: Modeling and Computational Approaches for Water Treatment
See more of this Group/Topical: Topical 1: Water Technology for Developed and Developing Countries (see also Separations Division)

(659e) Effect of Convective Hindrance On Microfiltration of Rod Shaped Particles

Ruth E. Baltus, Chemical and Biomolecular Engineering, Clarkson University, Potsdam, NY, Basavaraju Nagendra Agasanapura, Chemical & Biomolecular Engineering, Clarkson University, Potsdam, NY and Shankar Chellam, Department of Civil and Environmental Engineering, Department of Chemical Engineering, University of Houston, Houston

Microfiltraction and ultrafiltraction membranes are used in water and wastewater treatment facilities to produce biologically stable water. Efforts to develop theoretical models to quantify the *reflection coefficient* including hindrance and configurational effects on the particle transport date back several decades. These models have primarily focused on rejection of spherical particles [1]. However, numerous pathogenic microorganisms including bacteria and viruses are rod-shaped and it is not clear whether existing theories applicable for spheres can be directly applied to predict their transport and removal in membrane systems. The principal objective of our present study is to develop models describing transport of rod shaped particles across porous membrane.

We have previously reported experimental results for the rejection of several non-spherical bacteria and viruses from microfiltration membranes [2]. Results were compared to a theoretical model for capsule shaped particles that only considers configurational effects.

We have begun an effort to develop a more comprehensive model of particle transport that includes both (i) the steric constraints on the position and orientation of the particles and (ii) the convective hindrance (lag coefficient) experienced by rod shaped particles within cylindrical pores. As a first approximation, we have incorporated predictions for the lag coefficient for spherical particles into our previous model that includes steric restrictions for rod shaped particles. Predictions are made using different equivalent size parameters to describe the hydrodynamic resistance of the rod: the diameter of a sphere with equivalent volume, the rod length, and the rod diameter. Numerical results show that incorporating the lag coefficient for rejection coefficient predictions has a greater effect for particles with small aspect ratio (i.e, close to spherical shape).

We have also begun to perform CFD calculations to predict the lag coefficient for capsule shaped particles within a narrow channel with a rectangular cross-section. The objective here is to simulate fluid-particle interactions to evaluate the translation motion of the particle in a confined channel by calculating the hydrodynamic forces on the particle boundary[3]. The ratio of the translation velocity of the particle to the maximum fluid velocity yields the lag coefficient. The arbitary Lagrangian – Eulerian (ALE) algorithm is employed to simulate the system [3]. Results from both modeling approaches will be presented and discussed. The ALE method is verified with several previously reported standard benchmark problems.

[1] Dechadilok, P. and Deen, W.M., "Hindrance Factors for Diffusion and Convection in Pores," Industrial and Engineering Chemistry Research, 45, 6953-6959 (2006).

[2] Baltus, R.E., Xu, W., Badireddy, A.R. and Chellam, S. "Analysis of Configurational Effects on Hindered Convection of Non-spherical Bacteria and Viruses across Microfiltration Membranes," Industrial and Engineering Chemistry Research, 48, 2404-2413 (2009).

[3] S.M Davidson, K.V Sharp, "Boundary effect on the electrophoretic motion of cylindrical particles: concentrically and eccentrically positioned particles in a capillary" Journal of Colloid and Interface Science, 303, 288-297 (2006)

See more of this Session: Modeling and Computational Approaches for Water Treatment
See more of this Group/Topical: Topical 1: Water Technology for Developed and Developing Countries (see also Separations Division)

(691a) Fouling and Cleaning Behavior of Forward Osmosis Membrane (keynote)

Baoxia Mi, Civil and Environmental Engineering Department, The George Washington University, Washington, DC

Forward osmosis (FO) is considered a promising process in sea/brackish water desalination and wastewater reuse. FO membrane process uses natural osmotic pressure generated by a concentrated draw solution to pull water across a semi-permeable membrane from the feed solution. The diluted draw solution is then reconcentrated to recycle the draw solution as well as to produce purified water. Compared to conventional pressure-driven membrane processes (i.e. reverse osmosis/nanofiltration), FO process has the potential to reduce energy consumption and generate less brine discharge. In addition, our study found that the lack of hydraulic pressure in FO membrane operation is advantageous in terms of minimizing irreversible membrane fouling. Many types of membrane fouling (e.g., alginate fouling, gypsum scaling) in FO processes can be fully reversed by rinsing with pure water without using chemical cleaning reagents. This talk will focus on the fouling and cleaning behavior of single foulants (alginate, BSA, gypsum) as well as that of combined foulants in FO membrane processes. A microsope-equipped direct observation setup is used to monitor fouling layer formation on membrane surfaces through a glass window to characterize the kinetics of membrane fouling and cleaning. The adhesion force between foulant and membrane surface is determined by atomic force microscopy (AFM) to better understand the fouling mechanisms.

See more of this Session: Forward Osmosis for Water Treatment
See more of this Group/Topical: Topical 1: Water Technology for Developed and Developing Countries (see also Separations Division)

(691b) Permeation of Interacting Solutes In Forward Osmosis

William Phillip[1], Jui Shan Yong[2] and Menachem Elimelech[2], (1)Department of Chemical and Biomolecular Engineering, University of Notre Dame, Notre Dame, IN, (2)Chemical & Environmental Engineering, Yale University, New Haven, CT

Osmotically driven membrane processes are an emerging set of technologies that show promise in water and wastewater treatment, desalination, and power generation. The majority of research in the field has focused on predicting and enhancing water permeation through membranes, however, the effective operation of these systems requires that the reverse flux of draw solute from the draw solution into the feed solution be minimized. In this study, a model is developed that describes the reverse permeation of draw solution across an asymmetric membrane in forward osmosis operation. This model accounts for solute-solute interactions that can arise between solutes within both the active layer and support layer of the asymmetric membrane. Experiments were carried out to validate the model predictions. Using independently determined membrane transport coefficients, strong agreement between the model predictions and experimental results was observed.

See more of this Session: Forward Osmosis for Water Treatment
See more of this Group/Topical: Topical 1: Water Technology for Developed and Developing Countries (see also Separations Division)

(691c) Network-Based Finite Element Analysis for Studying the Effect of Support Structure On Internal Concentration Polarization During Forward Osmosis

Weiyi Li, Yiben Gao and Chuyang Y. Tang, Singapore Membrane Technology Centre and School of Civil and Environmental Engineering, Nanyang Technological University, Singapore, Singapore

Forward osmosis is one of the most promising technologies for water treatment with low energy requirement and energy generation (pressure retarded osmosis). A typical forward osmosis membrane is composed of a very thin active layer and a thick porous support layer. In contrast to the reverse osmosis, the forward osmosis process is mainly driven by the osmotic pressure difference between the draw solution and the feed solution. Previous studies found that severe flux decline could be caused due to the loss of effective driving force, which is extremely sensitive to the concentration profile developed within the porous support layer, namely the internal concentration polarization.

However, most of classical models accounting for the internal concentration polarization during forward osmosis are based on a linear approximation of the support structure while the complex geometry is neglected or lumped into the macroscopic phenomenological coefficients. Little work has been done to systematically study the underlying mechanisms accounting for the effects of the porous substructures on the coupled transport phenomena.

In the current work, the support structure was approximated by a three-dimensional network with stochastic disconnectivity, which was mathematically characterized by a blockage probability function. Then, a variety of membrane morphologies were simulated by varying the blockage probability in different coordinate directions. Both the bulk flow and the solute flow in the networks were numerically evaluated by using the finite element analysis, and therefore visualized to provide deeper insights into the interplay between the elaborate porous structures and the osmotically-driven convective diffusion. The presented modeling results also quantitatively revealed the profound correlations between the macroscopic transport parameters and the characteristics of the support substructures, thereby offering a very useful tool to optimize the support structure for controlling the internal concentration polarization during forward osmosis.

See more of this Session: Forward Osmosis for Water Treatment
See more of this Group/Topical: Topical 1: Water Technology for Developed and Developing Countries (see also Separations Division)

(691d) New Approaches to Forward Osmosis Membrane Design

Ngoc Bui[1], Jason Arena[1] and **Jeffrey R. McCutcheon**[2], (1)Chemical, Materials & Biomolecular Engineering, University of Connecticut, Storrs, CT, (2)Department of Chemical, Materials and Biomolecular Engineering, University of Connecticut, Storrs, CT

Forward osmosis (FO) and pressure retarded osmosis (PRO) rely on salinity gradients to drive water across a membrane. A major limitation to the widespread development and use of these technologies is the lack of an appropriately designed membrane. Today's thin film composite (TFC) membrane design consists of a three tiered structure consisting of a polyester (PET) nonwoven, an phase inversion precipitated polyethersulfone (PSu), and a polyamide selective layer. While this design has yielded highly permselective membranes for hydraulically driven separations, they exhibit poor performance during osmotically driven processes due to the low porosity and intrinsic hydrophobicity of the PSu layer that limits pore wetting. If the pore structure does not completely wet, the effective porosity of the support is reduced and internal concentration polarization is enhanced through a reduced solute diffusivity.

This paper will present two approaches to membrane design that aim to mitigate the impact of internal concentration polarization. First, we are able to modify commercial TFC membrane support layers using polydopamine which chemically alter the hydrophobic support layers for increased hydrophilicity. We are also building radically new TFC membrane support structures using electrospun nanofibers. Modification of TFC membranes has yielded a 5-12 fold increase in flux performance over the native membranes while our new membranes have exhibited a 2-5 fold improvement in flux over the best commercially available FO membranes.

See more of this Session: Forward Osmosis for Water Treatment
See more of this Group/Topical: Topical 1: Water Technology for Developed and Developing Countries (see also Separations Division)

(691e) Development of a Pilot-Scale Forward-Osmosis Desalination Process

Young Kim, Dong-Wook Oh, Yu-Chang Kim, Kong Hoon Lee and Jungho Lee, Energy Plant Research Division, Korea Institute of Machinery and Materials, Daejeon, South Korea

A pilot-scale forward-osmosis (FO) desalination system is designed to produce 3 ton per day of fresh water. In the front part of the system, fresh water is permeated from the feed, or seawater, to the draw solution in membrane modules. Subsequently, the draw solute is separated from water to produce the potable water in a membrane, or a column distillation system. The concentration of draw solution is designed with considering the capability of separation systems. The capacity of the separation systems is determined to enable treating the amount of draw solution required for target production of fresh water. The developed system is examined in performance tests. The results contribute to assessing the feasibility of forward-osmosis desalination systems.

See more of this Session: Forward Osmosis for Water Treatment
See more of this Group/Topical: Topical 1: Water Technology for Developed and Developing Countries (see also Separations Division)

(691f) Zero-Energy Osmotic Water Purifier for Ethiopia: Providing Clean Drink to Vulnerable Populations

Ethan Butler, Chemical, Materials & Biomolecular Engineering Dept., University of Connecticut, Storrs, CT and **Jeffrey R. McCutcheon**, Department of Chemical, Materials and Biomolecular Engineering, University of Connecticut, Storrs, CT

Globally, 894 million people do not have access to the basic amount of freshwater needed for drinking, cleaning and cooking. This is a catastrophe because water directly impacts every measure of development used by the Human Development Index (HDI): healthcare, education, and the economy. Many of the afflicted populations are also victims of natural disaster or migration. Refugees in particular are vulnerable populations that desperately need clean, sterile fluids for hydration and nutrition. Many water purification technologies are commercially available, but few are suitable for implementation in the rural of developing countries. Traditional methods of distillation, filtration, and disinfection require electrical inputs or chemical feed stocks that are difficult, if not impossible, to supply to rural areas. Some low tech options, such as sand filtration, offer an inexpensive solution to filtering large contaminants from water but do very little to remove harmful heavy metals and other dissolved contaminants.

This study investigates Ethiopia as a possible site for implementing a commercial forward osmosis process for producing a clean, nutritious drink from any fresh or brackish water surface water. *Forward osmosis (FO)* is an emerging technology that could be used to remove biological, organic, and inorganic contaminants from water. It works by utilizing an *osmotic pressure* gradient created by a "draw solution" to pull water via osmosis across a selectively permeable membrane while rejecting contaminants. For the specific process to be evaluated here, a sugar-nutrient-electrolyte solution mixture is used as the draw solution. This solution is diluted during osmosis, thus producing a clean, lightly sweetened, safe and nutritious drink. The concentrated contaminated water is returned to the source. Since this process occurs naturally, no energy input is required and since there is no exhaustive media (such as with adsorption), there is no hazardous waste associated with maintenance. The water is taken *directly* from a pond, stream or river with the concentrate being returned and diluted into the source.

The objective of this project was to assess the efficacy of a commercial forward osmosis system intended for use in rural Ethiopia and Ethiopian refugee camps. The particular system used during this study is the Hydration Technology Innovation's (HTI) Village System for its claimed field durability and capacity (200 liters of nutritious drink per day). The system employs a membrane capable of removing dissolved solids (organic and inorganic) and biological organisms (bacteria and viruses), though these performance metrics have yet to be tested and widely disseminated. In fact, no peer reviewed study has ever evaluated this system when functioning under real conditions. UCONN

Engineers without Borders has teamed up with HTI to evaluate their system performance under real conditions (in this case, purifying water from an on-campus lake). This study will show our preliminary pilot scale studies of using the HTI system hybridized with other low-tech options for pretreatment in order to preserve membrane life and performance. We will also present our findings from an assessment trip to Ethiopia to better understand how such a system can be implemented.

See more of this Session: Forward Osmosis for Water Treatment
See more of this Group/Topical: Topical 1: Water Technology for Developed and Developing Countries (see also Separations Division)

(762a) Desalination of the Colorado River Water: A Hybrid Approach

Anand Venkatesan and Phillip C. Wankat, School of Chemical Engineering, Purdue University, West Lafayette, IN

Reverse Osmosis (RO) is the most important technology for water desalination. The 96 million US gallons per day (gpd) Yuma desalination plant for desalination of irrigation return flows (3200 mg/L total dissolved salts (TDS)) in Colorado River (CR) is designed to operate at about 70% water recovery. Gabelich et al [1], studied RO desalination of CR water (887-1184 mg/L) in a pilot plant and demonstrated high recovery (≈95%). In this paper we design a hybrid system for desalination of CR water at the same capacity as the Yuma plant, for a feed with TDS = 941 mg/L.

The hybrid process comprised of Ion exchange (IEX) water softening and pH adjustment, integrated in between two RO steps. IEX treatment and pH adjustment of the concentrate from the Primary RO (PRO) is done before it is sent to the Secondary RO (SRO). IEX treatment of PRO concentrate helped remove Ca^{2+}, Mg^{2+} and Ba^{2+} which are the principal scale forming precursors and pH adjustment increased the solubility of Mg^{2+} salts. Use of concentrated retentate from SRO for regeneration of the IEX column eliminates the need for regeneration chemicals [2,3]. The hybrid process was evaluated for achieving high water recovery (≥95%).

The PRO and the SRO were modeled as cascades and optimal configurations were developed. Recovery in the PRO without the addition of antiscalants was 75%. The IEX was simulated with multiple columns feeding simultaneously and a single column regenerating. IEX treatment and pH adjustment of PRO concentrate allowed operation of SRO at 70% recovery and if antiscalants are added to control silica scaling SRO can be operated at recoveries as high as 95%. The overall recoveries were 92% without antiscalant addition and 98% with antiscalant addition in SRO. IEX/RO operation at higher recoveries reduces the amount of brine to be disposed. Economic analysis is done to study the impact of reduced brine on the cost of product water.

1. C. J. Gabelich et al., Journal of Membrane Science 301 (2007) 131-141
2. T. Vermeulen et al., Desalination, 47 (1983) 149-159
3. A. Venkatesan and P. C. Wankat, Desalination, 271 (2011) 122-131

See more of this Session: Water Reuse and Hybrid Water Treatment Processes
See more of this Group/Topical: Topical 1: Water Technology for Developed and Developing Countries (see also Separations Division)

TOWARD SUSTAINABILITY AND LOW COST IN ELECTRODIALYSIS REVERSAL DESALINATION

Maung Thein Myint, New Mexico State University, Las Cruces, NM 88003, mmyint@nmsu.edu, Abbas Ghassemi, New Mexico State University, Las Cruces, NM 88003, aghassem@nmsu.edu,

Summary

The improvements of electro-dialysis reversal (EDR) are discussed from the literature review. To be sustainable and low cost desalination, the areas of design and operation have to be *further* improved, and brine has to be reused in bio-energy production are suggested in details. Tertiary treated municipal wastewater (TTMWW) is proposed to use as feed in concentrate stream and electrode rinsing water in EDR to recover almost all of the fresh brackish groundwater (seawater) pumped for completely sustainable.

Key words: acid, chemical, electrode design, hydraulic leak, reusing concentrate from desalination; reusing tertiary treated MWW as concentrate in EDR.

Biography

Maung Thein Myint's Education

Yangon Technology University	*Civil Engineering*	*BE*	*1987*
NewMexico State University	*Environmental Engineering*	*MS*	*2001*
New Mexico State University	*Engineering*	*PhD*	*2005*
New Mexico State University	*Institute for Energy & the Environment; Post Doc*		*2009-present*

Abbas Ghassemi,

Education

University of Oklahoma	*Chemical Engineering*	*BS*	*1979*
New Mexico State University	*Chemical Engineering*	*MS*	*1989*
New Mexico State University	*Chemical Engineering*	*PhD*	*1990*

Appointments

Director, Institute for Energy & the Environment	*New Mexico State University*	*2006-present*
Professor, Chemical Engineering	*New Mexico State University*	*1999-present*
Executive Director, WERC	*New Mexico State University*	*1999-present*
Director, Institute for Energy & the Environment	*New Mexico State University*	*2008-present*

Introduction

Electro-dialysis reversal (EDR) desalination is known for its *excellent methods* to clean membrane. Membranes are cleaned by *reversing* the polarity between positive and negative polarity and by *switching* the hydraulic flow between concentrate and dilute streams in every fixed circle of polar reversal interval. Due to this excellence, researchers are trying to find ways to improve EDR in spacers, membrane's ability to withstand calcium sulfate, relaxing the fouling layer in membrane, and perm-selectivity of membrane. The spacer model was improved from Mark III to Mark IV to promote the turbulence and to increase the utilization area from 64 to 74%; to decrease power consumption from 0.14 to 0.10 kWh/m^3; and to increase the conductivity reduction from 27 to 36% (Grebenyuk and Grebenyuk, 2002). Acrylic based anion- and cation- exchange membranes that are highly resistance to irreversible resin fouling by organics, were introduced to use in 1981 and 1991 (Allison, 1995). The aliphatic anion selective membrane (AR 204 SXZL) ability to withstand sulfate fouling was improved to a saturation level of 440% $CaSO_4$ to gain a calculated water recovery rate (R_c) of 93.5% in a high level (42%) of SO_4^{2-} in feed water without pretreatment but with acid and anti-scalant additions in concentrate stream (Elyanow et al., 1981).

The costs of HCl and SHMP dosages in Dirab and Labakha-hawaita, Saudi Arabia are 15 and 83 times higher than power cost (Valcour, 1985). To avoid these high cost and un-sustainability with the chemical, EDR was successfully operated with low mean ion resident time in concentrate stream (*MIRTc*) by maintaining the similar LSI and $CaSO_4$ saturation level in the concentrate stream. For examples, EDR was successfully operated at LSI 2.29 and 358.9% $CaSO_4$ saturation level with the higher *R* 79.1% without adding any anti-scalants and without any acids in Turek et al., 2009 lab by *slowing down the velocity* in concentrate steam to gain the lower *MIRTc* and lower *dose* in single pass without any recirculation of concentrate from effluent into feed of concentrate stream. Moreover, Wisniewski et al., 2001 also demonstrated to operate ED without any recirculation to gain higher R 89.7% by *decreasing the volume* of concentrate for low dose and low *MIRTc*. The advantages of operating EDR in lower dose and lower *MIRTc* are to reduce the contact time between foulants and the surface of membrane in the concentrate steam, to reduce the dose to which is not high enough to be toxic to membrane, to increase the life time of membrane, and to eliminate chemical usage.

In the *classical* EDR, dimensions, flow, and velocity of dilute and concentrate are equal; LSI and $CaSO_4$ saturation level are used to control the scaling and fouling processes in concentrate (AWWA, 1995) as such LSI<+2.16 for preventing $CaCO_3$ from fouling and $CaSO_4$ saturation level<200 for averting $CaSO_4$ from precipitation. If LSI is more than allowable limit, acid is added in concentrate to keep $CaCO_3$ continuing dissolving (Katz, 1979); if $CaSO_4$ saturation level in concentrate is more than the allowable limit, antiscalant such as sodium hexametaphosphate (SHMP) is added in concentrate stream to maintain $CaSO_4$ enduring dissolving (Valcour, 1985). EDR however, was successfully *modernized* to operate with the higher water recovery rate (R) *without any anti-scalant and without acid*; this new EDR operated LSI at 2.29 and $CaSO_4$ saturation level 358.9% at *lower dose and lower* $MIRT_c$ (Turek et al., 2009). Dose and $MIRT_c$ are proposed to address the controlling process. By lowering *R* and polar reversal interval, EDR can be operated at MIRTc<130 min; at MIRTc<130min, desalting cost/energy can be minimized by eliminating chemicals requirement (Myint et al., 2010a, 2011a, and 2011b).

Brine disposal cost could range from 5 to 33% (Khordagui, 1997; Mohamed et al., 2005). In the case of inland sites, this minimum cost increases to the order of 15% of the costs of desalination (Glueckstern and Priel, 1996; Oren, et al., 2010). The cost for concentrate disposal by evaporation pond is 1.18-10.04 $\$/m^3$ (Greenlee et al., 2009). At present, all the best available disposal methods are highly questionable to the environmental concerns, and therefore everywhere in literature (i.e., Myint et al., 2010b) called for the recovery the salt or reusing of brine concentrate wasted from desalination in bioenergy and microalgae biodiesel production to be *sustainable* and to *dramatically reduce* the cost.

Suggestion

Water leaked significantly from the membranes, due to sheet membranes are stacked together, in higher R and higher demineralization rate as shown in Elyanow et al., 1981 that the metered water recovery rate (R_m) 86.0% was different from calculated R_c (93.5%). Although membrane has the ability to withstand the saturation level of $CaSO_4$ to 440%, the set objective was not achieved due to the hydraulic leak. The Research in reverse osmosis stated that acid and anti-scalant addition hinders the permeability rate (Hasson et al., 2007). To gain the higher R_c (equal R_m) with lower desalination cost, researches in hydraulic leaks and acids and antiscalant additions have to be revised in EDR.

Moreover, ED/EDR is generally assembled with 100 cell pairs for low salinity feed water to 700 cell pair for the high salinity brackish water (Jain and Reed, 1985) in one electrical stage due to the current transferring area available in electrodes. The positive and negative electrodes that have the same sizes of the membrane are usually placed in the top and bottom of the membrane cell pairs. The feed water is supplied into influent of the cell pair and collected from the effluent of the cell pairs. The TDS concentration gradually decreases from influent to effluent due to the voltage (direct current) supplied in the electrodes. The constant voltage is typically supplied to the electrodes uniformly along the surface. The current is carried by ionic molar flux (Kim et al., 2011); the current near the effluent is small due to the less TDS which leads to larger resistance ($R = V/I$) in the constant voltage potential application case. The electrode design has to be updated with the higher current intensity in influent and lower current intensity in effluent gradually along with the TDS concentration profile.

One way to be sustainable is to reuse tertiary treated municipal wastewater (TTMWW) as feed in concentrate stream and rinsing water in electrode channels. TTMWW has been successfully used as makeup in cooling system of cooling tower of coal-fired power plant such as Redhawk Power Plant in Arizona which uses TTMWW for over 90% of the cooling system makeup (Li et al., 2011). The trivial cost for tertiary treatment is much lower than the cost paid for dam water or desalinated water (Papaiacovou, 2001). *EDR membrane* has demonstrated to handle continuous microfiltration pretreated MWW effluent from the EDAR del Sureste to produce the high quality of the treated water pilot plant studies (Pino and Durham, 1999) and field scale in Canary Islands (Veza, 2001). Other successful story of field scale EDR membrane desalting TTMWW are 300,000 GPD in Moody Gardens in Galveston, Texas and two of 1 MGD in Middle East without organic fouling and biogrowth, however, with colloidal fouling in half of the plants (Allison, 1995).

The ultrafiltration (UF) with module ZW-10 has capability to in the removal of *E. coli* from secondary treated MWW (Table 1). In the case of *E. coli*, in particular, UF removed 100% of *E. coli* (Lubello et al., 2003). The barriers also proved capable, although not absolute, in reducing TC as was expected. UF with module ZW-10 reduces turbidity (94.5%), TSS (98.7%), and COD (around 35%) from secondary treated MWW (Table 2) and produces permeate with low SDI values (usually < 3%) (Lubello et al., 2003). By using UF treated TTMWW as feed into concentrate stream and electrode rinsing water, therefore, all of the freshwater may be desalted into a drinking water, and all the TDS will be collected in concentrate stream in TTMWW. This idea will result in overall higher fresh water recovery rate.

UF with module ZW-10 removes 100% of the *E. coli*; dilute water rinses out all the concentrate (TTMWW) which attaches in surface of membrane from previous cycle for 30 seconds before product water produce in EDR (AWWA, 1995); dilute product water does not directly contact with TTMWW. Although, reusing TTMWW as feed in concentrate and electrode rinsing waters in EDR which produces drinking after the *E. coli* treatment in dilute product water, is technically feasible and completely sustainable, perception of the public may be problem.

Conclusion

With the updating design in electrodes; improving in hydraulic leaks; reducing in chemical usage in concentrate stream; reusing TTMWW as feed in concentrate steam and electrode rinsing water; and reusing brine in bio-energy production; EDR will have a capability to produce almost 100% of water recovery rate from fresh brackish groundwater (seawater) pumped with the low desalination cost.

Acknowledgement

We thank and appreciate the funding agency the Office of Naval Research (Contract # N00014-08-1-0304) from the USA and Tier I: Reclamation Cooperative Agreement Program.

Reference

1. Allison, R. P. (1995) Electrodialysis reversal in water reuse applications. *Desalination* **103**, 11-18.
2. AWWA (1995) Manual of Water Supply Practices- M38, First Edition: Electro-dialysis and Electro-dialysis Reversal. American Water Works Association, Denver, CO.,.
3. Elyanow, D., Sieveka, E., and Mahoney, J. (1981) The determination of super-saturation limits in an EDR unit with aliphatic anion membranes. Ninth Annual Conference and International Trade Fair of the National Water Supply Improvement Association, Volume II. May 3-June 4.
4. Glueckstern, P. and Priel, M. (1996) *Desalination* **108**, 19.
5. Grebenyuk, V. D. and Grebenyuk, O. V. (2002) Electrodialysis: From an Idea to Realization. *Russian Journal of Electrocchemistry* **38**(8) 806-809.
6. Greenlee, L. F., Lawler, D. F., Freeman, B. D., Marrot, B., Moulin, P. (2009) Reverse osmosis desalination: water sources, technology, and today’s challenges. *Water Research* **43**, 2317-2348.
7. Hasson, D., Drak, A., Komlos, C., Yang, Q., Semiat, R. (2007) Detection of fouling on RO modules by residence time distribution analyses. *Desalination* **204**, 132–144.

8. Jain, S. M. and Reed, P. B. Electro-dialysis. In *Comprehensive Biotechnology: The Principles, Applications, and Regulations of Biotechnology in Industry, Agriculture and Medicine, Volume 2: The principles of Biotechnology: Engineering Considerations*; Moo-Young, M., Cooney, C. L., Humphrey, A. E., Eds.; Pergamon Press: Oxford 1985; pp 575-590.
9. Katz, W.E. (1979) The electro-dialysis reversal (EDR) process. *Desalination* **28**, 31-40.
10. Khordagui, H. (1997) Environmental Aspects of Brine Reject from Desalination Industry in ESCWA Region. ESCWA, Beirut, Lebanon.
11. Kim, Y., Walker, W. S., and Lawler, D. F. (2011) Electrodialysis with spacers: effects of variation and correlation of boundary layer thickness. *Desalination* **274**, 54-63.
12. Li, H., Chien, S-H., Hsieh, M-K., Dzombak, D. A., and Vidic, R. D. (2011) Escalating Water Demands for Water Production and the Potential for Reuse of Treated Municipal Wastewater. *Environ. Sci. Technol.*, **45**, 4195–4200.
13. Lubello, C., Gori, R., de Bernardinis, A. M., and Simonelli, G. (2003) Ultrafiltration as tertiary treatment for industrial reuse. *Water Science and Technology: Water Supply* **3**(4) 161-168.
14. Mohamed, A. M. O., Maraqa, M. and Handhaly, J. A. (2005) Impact of land disposal of reject brine from desalination plants on soil and groundwater. *Desalination* **182**, 411-433.
15. Myint, M. T., Ghassemi, A. and Nirmalakhanda N. (2010a) Design of An ILEDR for Brackish Groundwater: A Literature Review Approach. *Desalination and Water Treatment* **24,** 150–175.
16. Myint, M. T., Ghassemi, A. and Nirmalakhanda N. (2010b) Review of Nondestructed Anaerobic Digested Sludge to Reinstate Concentrate from Electrodialysis Reversal Desalination in Bioenergy Production. *Environmental Engineering Science* **27**(10) 811-823.
17. Myint, M. T., Ghassemi, A. and Nirmalakhanda N. (2011a) Low Energy Desalination: Low Dose & Low Mean Ion Resident Time in Concentrate of Electrodialysis Reversal. 7th IWA leading-edge conference on water and wastewater technologies 2010Phoenix, Arizona, USA. *Water Science Technology* **63**(9) 1855-1863.
18. Myint, M. T., Ghassemi, A. and Nirmalakhanda N. (2011b) Modeling in Desalination: Electro-dialysis Reversal. *Desalination and Water Treatment* **27**, 255-267.
19. Oren, Y., Korngold, E., Daltrophe, N., Messalem, R., Volkman, Y., Aronov, L., Weismann, M., Bouriakov, N., Glueckstern, P., Gilron, J. (2010) Pilot studies on high recovery BWRO-EDR for near zero liquid discharge approach. *Desalination* **261**, 321–330.
20. Papaiacovou, I. (2001) Case study — wastewater reuse in Limassol as an alternative water source. *Desalination* **138**(1-3) 55-59.
21. Pino and Durham, 1999 Wastewater reuse through dual-membrane processes: opportunities for sustainable water resources. *Desalination* **124**(1-3) 271-277.
22. Turek, M., Was, J., and Dydo, P. (2009) Brackish water desalination in RO- single pass EDR system. *Desalination and WaterTreatment* **7**: 263-266.
23. Veza, J. M. (2001) Desalination in the Canary Islands: an update. *Desalination* **133,** *259-270.*
24. Wisniewski, C., Persin, F., Cherif, T., Sandeaux, R., Grasmick, A., and Gavach, C. (2001)

De-nitrification of drinking water by the association of an electrodialysis process and a membrane bioreactor: feasible and application. *Desalination* **139**, 199-205.
25. Valcour Jr, H.C. (1985) Recent applications of EDR. Conference Proceedings: Second World Congress on Desalination and Water Re-use. Bermuda, Nov 17-22.

Table **1**. Results of microbiological analysis treated by UF with ZW-10

n.	TC (MPN/100 ml)		E. Coli (UFC/100 ml)	
	UF concentrate	UF permeate	UF concentrate	UF permeate
1	380,000	1,950	29,000	0
2	93,000	11,000	1,900	0
3	93,000	11,000	1,000	0

Lubello et al., 2003.

Table **2**. Quality of feed, permeate and concentrate treated by UF with ZW-10

Parameters	Feed (TTMWW)			Permeate from UF			Concentrate from UF		
	Mean	Range	STD	Mean	Range	STD	Mean	Range	STD
COD (mg/L)	41.4	27–68	10.2	27.4	17–53	8.2	77.6	34–116	22.7
Turbidity (NTU)	7.35	2.2–32	6.9	0.4	0.18–0.70	0.18	29.9	7.7–70	16.2
TSS (mg/L)	23.2	3.7–71.1	15.9	0.32	0.71	0.18	45.5	15–74.8	16.7

Lubello et al., 2003

"Municipal wastewater treatment using electrocoagulation and adsorption for reuse in industrial processes and green areas"

Hector Moreno, Alejandro Romero, Jose R. Parga, Jewel Andrew Gomes, Eric Peterson.

One of the biggest challenges of the 21st century is to provide water, primarily potable, to the majority of the world population. This is a critical issue, especially in developing countries. One way to augment that effort is to reuse water. Currently, some emerging wastewater treatment technologies based on electrochemistry are available that are competitive and advantageous over conventional technologies (e.g., Electrocoagulation (EC), electro-flotation, electro-decantation and electro-oxidation). Although EC has been known for more than a century, it has not been comprehensively studied. EC has been commercialized for the removal of specific contaminants and wastewater treatment. This work demonstrates how EC has been studied experimentally and is proposed as an alternative method for the biological treatment of municipal wastewater. Experiments were conducted to determine the optimum operational conditions. Results show a remarkable removal efficiency for: chemical oxygen demand (COD) 77 to 94%; for coliforms 80%, and; colony forming units (CFU) 99.98%; within 30 seconds residence time.

Key words: Electrocoagulation, COD, CFU, municipal wastewater.

Introduction

Wastewater is the liquid end-product (by-product), of municipal, agricultural, and industrial activity. The chemical composition of wastewater reflects the origin from which it came. The term wastewater implies it is a waste product to be discarded. However the water we use (e.g. drinking, washing, bathing, etc.) ultimately ends up back in the stream, river, lake, or groundwater to be withdrawn, treated, and used again. So wastewater must be treated and/or reused in an environmentally sound manner before it is discarded [1].

Water pollutants consist of floating materials, settable solids, organic and inorganic suspended solids, organic and inorganic dissolved solids and liquids, oil, fats and grease, dissolved gases and microorganisms. All water pollutants have different characteristics, poseing separate threats for human health and their removal.
Wastewater treatment plants are devoted to remove most of the pollutants while delivering water with little or no risk back in the stream, river, lake, groundwater or industrial reuse.

Conventional wastewater treatment include: screening, coagulation – flocculation, sedimentation, biological treatment, filtration and disinfection.

Table 1 show typical concentrations of some key parameters that determine the efficiency of wastewater treatment and what will be its usage after the treatment, table 1 shows the typical concentrations. Fecal coliforms and CFU are especially important in removing because communicable diseases can be transmitted by the pathogenic organisms in wastewater.

Table 1. Typical contaminant concentration of untreated domestic wastewater (After Metcalf and Eddy (1991).

		Concentration		
Contaminants	**Unit**	**Weak**	**Medium**	**Strong**
BOD_5 at 20° C	mg L^{-1}	110	220	400
Chemical oxygen demand (COD)	mg L^{-1}	250	500	1000
Grease	mg L^{-1}	50	100	150
Total coliform	CFU 100 mL^{-1}	10^6-10^7	10^7-10^8	10^8-10^9
Total coliform	number per milliliter	10^4		10^5

Biological Oxygen Demand (BOD), is defined as the amount of oxygen required by bacteria to decompose organic matter for a specified period of time (usually 5 days), under aerobic conditions. The amount of oxygen reported with this method represents only the carbonaceous oxygen demand (CBOD) or the easily decomposed organic matter. BOD is commonly used to measure natural organic pollution. Its analysis technique uses a titrametric determination of O_2 consumption due to the biological oxidation of sample.

An alternative to the BOD test for determining the oxygen consuming potential of a wastewater sample is the chemical oxygen demand (COD) test. As the name implies, carbonaceous oxygen demand is oxidized chemically in the COD test. Like BOD, the units for COD are in milligrams of oxygen per liter (mg L^{-1}).

Chemical Oxygen Demand (COD), is the amount of oxygen required to oxidize organic and oxidizable inorganic matter present in water or wastewater. Its analysis technique determines oxidant consumed during a chromic acid digestion of the sample in standard conditions.

The advantage of this test is that it is quick and reproducible. Analysis of COD is much more practical than analyses for specific pollutants.

A disadvantage is that it measures everything, including non-biodegradable organic matter, that can be oxidized by potassium dichromate. There are also some incompatibilities with the COD test. For example, low molecular weight fatty acids and aromatic hydrocarbons may not be well oxidized during the test while inorganic ions (e.g. chloride and nitrite for instance) may be oxidized. The addition of certain catalysts during the test can eliminate most of these interferences [2].

EC Removal efficiency

There is an extremely high variability in organic compound removal efficiencies for organic compounds, BOD and COD using EC. This is one reason why EC is not universally accepted as a conventional technology. Removal efficiency for COD might show negative, medium and surprising high values. It all depends on the compounds that contribute to COD and their reactivity with iron. For example suspended solids in small quantities are not a problem for EC, organic compounds reacting with iron (II) and iron (III) to form insoluble compounds are also removed. Organic compounds reacting with iron (II) to form insoluble compounds and with iron (III) to form soluble compounds or inversely will be partially removed. Organic compounds that do not react with iron (II) or iron (III) will not be affected by the EC process [3].

Current theory of EC involve several successive stages:

1) ***Generation of metal ions.***

Anode:

$$Fe \rightarrow Fe^{+2} + 2e^{-} \quad (1)$$

$$Fe \rightarrow Fe^{+3} + 3e^{-} \quad (2)$$

2) ***Hydrolysis of metal ions and generation of metal hydroxides***

$$Fe + 6H_2O \rightarrow Fe(H_2O)_4(OH)_{2\,(aq)} + 2H^{+1} + 2e^{-1} \quad (3)$$

$$Fe + 6H_2O \rightarrow Fe(H_2O)_3(OH)_{3\,(aq)} + 3H^{+1} + 3e^{-1} \quad (4)$$

3) ***With iron electrodes, Hydrogen evolution at the cathode.***

Cathode:

$$2H^{+} + 2\,e^{-} \rightarrow H_{2(g)}\uparrow \quad (5)$$

Comments:

Electrochemistry depends on thermodynamics and kinetics. The rate of reaction depends on the dissolution of iron and the removal of $[H^{+}]$ via H_2 evolution. This reaction occurs fast for low pH values (strong acids) and for a weak acid the rate will depend on the pKa of the acid.

4) ***Destabilization of the contaminants, particulate suspension, breaking of emulsions, and aggregation of the destabilized phases to form flocs.***

This part relative to colloids and suspended matter is accepted since in small quantities they are not a problem for EC.

5) ***Theory of EC consider that chemical reactions and precipitation can occur during the EC process or that other cation and/or the hydroxyl ion (OH) form a precipitate with the pollutant.***

EC can be considered an accelerated corrosion process with Green Rust (GR) recognized as an important intermediate phase in corrosion of Fe (0).

Green rust (GR). GRs are layered Fe(II)-Fe(III) hydroxides having a pyroaurite-type structure consisting of alternating positively charged hydroxide and hydrated anion layers. Some Fe (II) of the octahedral sheets of $Fe(OH)_2$ is replaced by Fe(III). This results in positive layers of charge that are balanced by the inclusion of anions between the layers. Its exact nature depends on the interlayer anion. Various forms have been synthesized and studied. Results of many studies show GR conforming to a general chemical composition and stoichiometry, that can be represented with the general formula: $[Fe^{II}_{(6-x)}\ Fe^{III}_{x}(OH)_{12}]^{x+}[(A)_{x/n}\cdot yH_2O]^{x-}$, where x ranges from 0.9 to 4.2, A is an n-valent anion (typically CO_3^{2-}, Cl^- or SO_4^{2-} and y denotes the varying amounts of interlayer water (typically y ranges from 2 to 4 for most GRs) [4].

GR's, unlike most iron oxides and their form, have great specific internal surface areas and consequently great reactivities. They are reactive ion exchangers and sorbents (Hansen et al., 2001). The amphoteric surface hydroxyl groups lead to both sorption of heavy metals (cations) and organic anions, e.g. linear alkyl benzene sulfonates (LAS), the major synthetic surfactant in laundry detergents and cleaning products, and inorganic anions, e.g. silicate, arsenate and selenate. Polar non-charged compounds are sorbed into the interlayer. Cations (e.g. Cu(II), Ni(II), Zn(II), Cd(II), Co(II) and Mg(II)) may isomorphically substitute for Fe(II) during GR formation (co precipitation). For instance, nickelous-ferric GR, end products obtained by substitution of Fe(II) ions by Ni(II) ions in GR (Refait & Génin (1993, 1997) and Refait et al. (1994, 1998)). This incorporation of cationic inorganic contaminants into the structures of GR may provide an effective means of sequestering contaminants in the subsurface.

The reactivity of GR is responsible of the removal of cationic and anionic pollutants, by substituting atoms of Fe (II) or (III) or OH radicals.

Overall reaction. In general, hydrogen gas and GR are formed at the cathode as shown in equation (6) [5]:

$$6Fe + (12+x)\ H_2O \rightarrow \frac{1}{2}\,(12+x)H_{2(g)}\uparrow + x\ Fe(OH)_3 * (6\text{-}X)\ Fe(OH)_{2\,(s)} \qquad (6)$$

6) ***Formation of rust (dehydrated hydroxides) occurs as shown in the following:***

$$2Fe(OH)_3 \leftrightarrow Fe_2O_3 + 3H_2O \qquad \text{(hematite, maghemite)} \qquad (7)$$

$$Fe(OH)_2 \leftrightarrow FeO + H_2O \qquad (8)$$

$$2Fe(OH)_3 + Fe(OH)_2 \leftrightarrow Fe_3O_4 + 4H_2O \qquad \text{(magnetite)} \qquad (9)$$

$$Fe(OH)_3 \leftrightarrow FeO(OH) + H_2O \qquad \text{(goethite, lepidocrocite)} \qquad (10)$$

Hematite, maghemite, rust, magnetite, lepidocrocite and goethite have been identified as EC by-products.

Experimental

The corresponding experiments, used the following: Lab materials, beakers of different capacities, iron electrodes of 59x32x2 mm placed 6 mm apart, Steren MUL-285 multimeters, a model ELI-055 Steren voltage converter 110 ac V – 15 dc V, an Arrow 82608, 0.56 KW four poles interruptor to reverse current, a LabTech Hot plate stirrer; Model LMS-1003, for filtration, 150 mm Whatman # 1 filter papers, an HI/839800 Hanna Digestor, COD Hanna vials and an HI83099 Hanna Photometer for COD determination.

A picture of the experimental setup is shown in figure 1. The wastewater sample is in the beaker size rector; as electrodes, iron plates are used and current flow is reversed every minute if required with a four pole interrupter, to avoid electrode polarization and uneven plate consumption. Plates are connected to an amperimeter and a voltmeter to measure current and voltage.

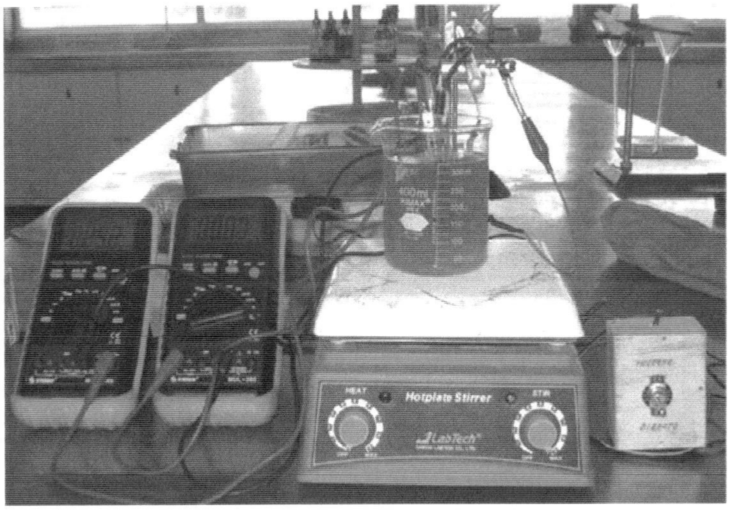

Figure 1. Experimental setup

All samples were taken from the "Las Fuentes" collector, next to the El Campestre wastewater treatment plant in Torreon; Coahuila, Mexico, and were run depicted in the process flow diagram shown as Figure 2. A sample was also treated with an adsorbent agent (WAA). Three replicates were made with the medium value reported.

Figure 2. Process flow diagram of the experimental tests.

MPN and CFU were determined using EPA standard methods in the Biology Faculty of the Coahuila Autonomus University, Campus Torreon.

RESULTS AND DISCUSSION

Experiment 1: run to assess the quality of the method, on sample 1, for 5 minutes. Parameters and results are shown in Table1.

Table 1. Parameters and results for experiment 1.

	V	A	Conductivity (μS)		pH		Initial COD (mg/lt)	Final COD (mg/lt)	% COD Removal
			Initial	Final	Initial	Final			
	12.58	0.54	923	764	9.07	9.39	245	122	50.24
WAA	-12.96	-0.49		902		9.47		81	66.93

Experiment 2: five minutes residence time showed good results, with experiment run for ten minutes to test the effect of residence time. Parameters and results for this test are presented in Table 2.

Table 2. Parameters and results for experiment 2.

	V	A	Final Conductivity (μS)	Final pH	Initial COD (mg/lt)	Final COD (mg/lt)	COD Removal (%)
			889	9.59	245	123	49.79
WAA	13.4	0.47	842	9.56		74	69.79

Results are similar to those for experiment 1, so it is advisable to run EC for five minutes, but stopping the process every minute and determining COD removal efficiency.

Experiment 3: to determine the effect of residence time on COD removal efficiency. Results can be seen in Table 3 and Figure 3.

The best result, adsorbent agent added, was attained within one minute of residence time, with little difference between trials.

Some samples turned yellowish red, which indicating the presence of excess iron and its oxidation.

Table 3. Determination of the effect of residence time on COD removal (Exp 3)

Residence time t_r (min)	V	A		Conductivity (μS)	pH	Initial COD (mg/lt)	Final COD (mg/lt)	COD Removal (%)
1	13.01	0.53		859	9.2		144	41.22
			WAA	917	9.55		82	66.53
2	15.03	0.46		850	9.21		141	42.44
			WAA	908	9.75		84	65.71
3	14	0.46		926	9.21		121	50.61
			WAA	834	9.77	245	98	60
4	13.24	0.48		908	9.27		136	44.49
			WAA	804	9.91		104	57.55
5	13.23	0.47		912	9.3		127	48.16
			WAA	833	9.68		86	64.89

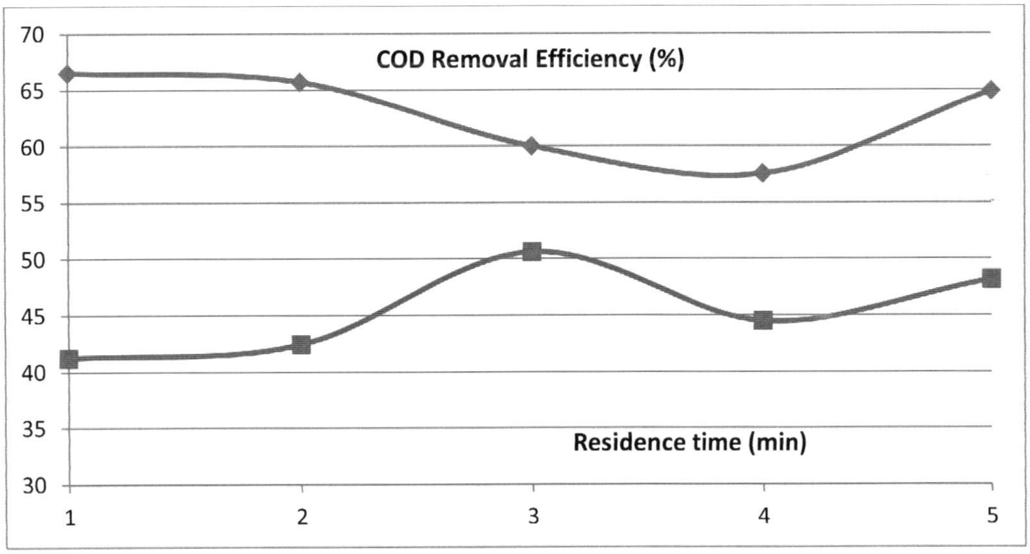

Figure 3. Graphic of COD removal efficiency vs. Residence time for experiment 3.

Experiment 4: To continue residence time on COD removal efficiency, samples were run at 15, 30 and 45 seconds of residence time. Results are seen in Table 4, and Figures 4 - 7 show the process. In figure 4, green rust is seen after EC. Figure 5, shows that after filtration, some color and odor are still present. Figure 6, reveals that after adding adsorbent agent and filtration, color and odor are gone.

Table 4. Determination of the effect of residence time on COD removal (Exp 4)

Residence time t_r (sec)	V	A	Watts		Conductivity (μS)	pH	Initial COD mg/lt)	Final COD (mg/lt)	COD Removal (%)
15	11.22	0.62	6.95		926	8.56	245	69	71.83
				WAA	1057	9.59		25	89.79
30	12.08	0.57	6.88		910	8.67		61	75.10
				WAA	1003	10.02		14	94.28
45	11.02	0.6	6.61		923	8.75		65	73.46
				WAA	984	10.04		46	81.22

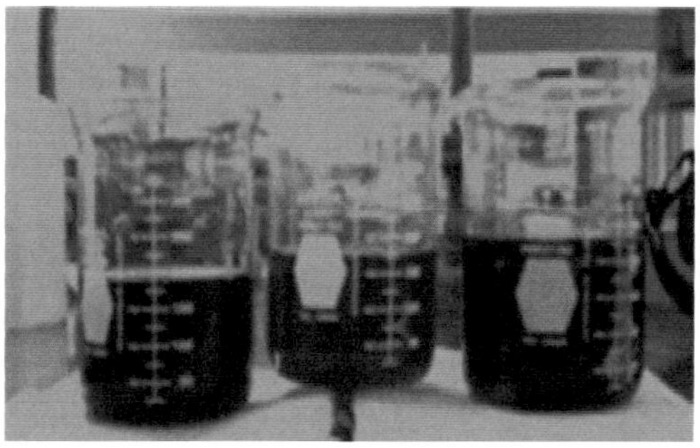

Figure 4. Treated wastewater after EC at 15, 30 and 45 sec (Experiment 4)

Figure 5. Treated wastewater after filtration, from left to right 15, 30 and 45 sec

Figure 6. Treated wastewater after addition of adsorbent agent and filtration , from left to right 15, 30 and 45 sec

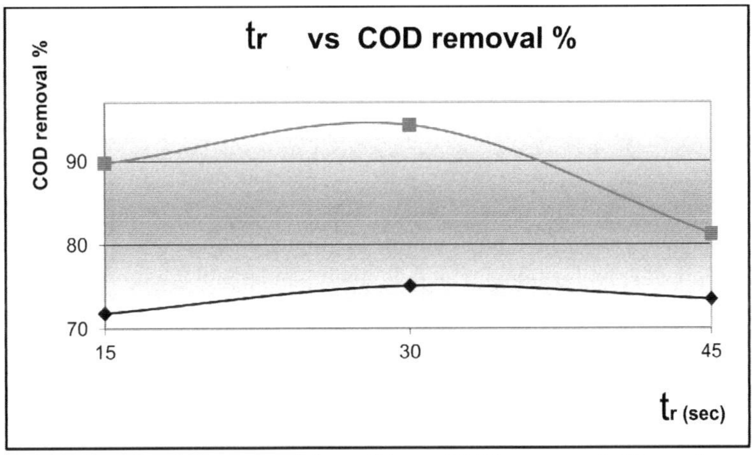

Figure 7. Graph of Removal time (t_r) vs COD removal, (Experiment 4)

Less than a minute was needed to completely treat wastewater. A COD removal efficiency of 94.28 % was achieved within 30 seconds using the adsorbent agent. With these parameters the energy consumption is just 7 watts.

Experiment 5: Three additional samples run at a residence time of 30 seconds with results presented in tables 5 - 7.

Table 5. COD removal with 30 seconds of residence time (Exp 5.1).

Residence time t_r (sec)	**V**	**A**		**Conductivity (μS)**	**pH**	**Initial COD (mg/lt)**	**Final COD (mg/lt)**	**COD Removal (%)**
30	11.18	0.58		950	8.01	352	187	46.87
			WAA	1000	7.63		80	77.27
30	10.2	0.59		940	8.5	282	122	56.73
			WAA	1000	9.0		65	76.95
30	10.2	0.59		900	7.3	342	92	73.09
			WAA	860	7.27		51	85.08
Average								58.89
			WAA					79.77

Results are consistent with previous ones presented results and with those obtained for biological wastewater treatment plants.

Experiment 6: is designed to determine the removal efficiency of MPN of coliforms in wastewater with EC. A sample was run using the residence time determined in previous experiments. Results are presented in table 6.

Table 6. EC parameters, Initial and final MPN. (Experiment 6)

Residence time t_r (min)	**V**	**A**		**Conductivity (µS)**	**pH**	**Coliforms MPN**		**Coliforms Removal (%)**
						Initial	**Final**	
30	11.64	0.56		1006	7.42	>2400	1100	54.16
			WAA	985	7.28		480	80

Experiment 7: examines the EC process for CFU, again run for 30 seconds with results shown in Table 7.

Table 7. Parámetros iniciales y finales de la EC de la (Prueba 6)

Residence time t_r (min)		**Conductivity (µS)**	**pH**	**CFU/ml**		**CFU Removal (%)**
				Initial	**Final**	
30		900	7.30	2,130,000	21,000	99.01
	WAA	860	7.27		400	99.98

The attained results are excellent as the removal efficiency for CFU is almost 100%.

CONCLUSIONS

It has is demonstrated that using EC in municipal wastewater treatment is efficient and effective for removing COD, Coliforms and CFU. Residence times less than only a minute are required to complete the wastewater treatment, with 30 sec established as an optimal residence time. COD removal is good and comparable to oxidation pond biological treatment where residence times range from 3 to 6 days. Conventional methods, (e.g. biological) also generate odors and a lot of sludges. EC generates just the exact required amount of coagulant. EC wastewater treatment might substitute primary and secondary treatment in conventional treatment plants due versatility and substantially lower residence times.

According to the data seen in table 8, for a residence times between 1 to 10 minutes, COD removal is almost constant (64 % WAA). With residence times of 15 to 45 seconds the mean value for COD removal is 84% with a variance of 48 which makes the results consistent.

For all residence times COD removal is higher WAA.

Table 8 . Comparison of COD removal for various residence times.

Experiment	**Residence time t_r (min)**	**COD removal (%)**		**Experiment**	**Residence time t_r (sec)**	**COD removal (%)**	
			WAA				**WAA**
1	5	50.24	66.93	4	15	71.83	89.79
2	10	49.79	68.79		30	75.1	94.28
3	1	41.22	66.53		45	73.46	81.22
	2	42.44	65.71	5.1	30	73.09	85.08
	3	50.61	60.0	5.2	30	46.87	77.27
	4	44.49	57.55	5.3	30	56.73	76.95
	5	48.16	64.89				
Mean (μ)		46.71	64.06	Mean (μ)		66.18	84.10
Variance (σ^2)		15.43	14.01	Variance (σ^2)		134.88	48.61

The maximum permissible values of pollutants allowed for the official Mexican norm NOM-003-ECOL-1997, for residual treated wastewaters for reuse in services to the public are presented in table 9

Table 9. NOM-003-ECOL-1997

Type of reuse	Monthly average				
	Fecales Coliforms MPN/100 ml	Helminto eggs (e/l)	Oil & grease mg/lt	BOD_5	TSS mg/lt
Service to the public with direct contact.	240	1	15	20	20
Service to the public with occasional or indirect contact.	1000	5	15	30	30

Comparison with an original simple coliform removal efficiency of 80%, (e.g. > 2400 MPN of coliformes) a minimum 480 MPN of coliforms is attained while the norm indicates a maximum value of 1000 for services with indirect or occasional contact to the public. Results comply with the regulations at less than half the maximum value allowed, and treated wastewater may be used for reuse in green areas and industrial process as well.

Objective are accomplished, investigate the application of EC combined with the use of an adsorbent agent for wastewater treatment and show that it is possible to determine parameters for optimum operation, and foundation for industrial application in a continuous process were established.

Acknowledgements

The authors wish to acknowledge the support for this project to the National Council of Science and Technology (CONACYT) and to the Dirección General de Educación Superior Tecnológica (DGEST) from Mexico.

[1] WASTEWATER GENERATION & TREATMENT: DOMESTIC SEWAGE , Published By : Dr. B. Sengupta, Member Secretary, Central Pollution Control Board, Delhi – 32, at ENVIS Centre – 01, Text : Dr. R.C. Trivedi, Sh. R.M. Bhardwaj, Dr. S. Agrawal, Sh. Nazimuddin, Ms. Kokil Malhotra and Sh. Mahendra Pandey]

[2] Environmental and Ecological Chemistry, Edited by : Aleksandar Sabljic, Rudjer Boskovic Institute, Croatia, Part 4. CHEMISTRY OF WASTEWATER, Timothy G. Ellis, Department of Civil, Construction and Environmental Engineering, Iowa State University, Ames, Iowa, USA 2004.

[3] Hector A. Moreno-Casillas , David L. Cocke , Jewel A.G. Gomes, Paul Morkovsky , J.R. Parga, Eric Peterson ; **Electrocoagulation mechanism for COD removal**; Separation and Purification Technology 56 **(2007)** 204–211

[4] G. Chavez Lucia H. "The role of green rust in the environment: a review" Revista Brasileira de Engenharia Agricola e Ambiental, v9, n2, 284-288, 2005.

[5] H. A. Moreno C, D. L. Cocke, J. A. G. Gomes, P. Morkovsky, J. R. Parga, E. Peterson and C. Garcia; Electrochemical Generation of Green Rust using Electrocoagulation; ECS Transactions, 3 (18) 67-78 (2007 10.1149/1.2753225, The Electrochemical Society

(762d) Transport and Remediation of Chromium (VI) In Groundwater Using An Integrated Electrokinetic and Nanoscale Particle Technology

Charlotte Chan, **Ryan Thacher**, Lewis Hsu and Massoud Pirbazari, Sonny Astani Department of Civil and Environmental Engineering, University of Southern California, Los Angeles, CA

Hexavalent chromium [Cr(VI)] is one of the EPA's target contaminants due its prevalence in natural waters in concentrations above levels regarded as safe. When present in soils of low hydraulic conductivity such as clay, conventional treatment technologies become ineffective due to the difficulty and energy consumption in pumping through tightly packed soil. An emerging novel technology, electrokinetic remediation, uses electric potential to transport ionic contaminants through saturated soils and groundwater, eliminating problematic issues associated with state-of-the-art remediation technologies. Strategic placement of electrodes surrounding a contaminated zone with low voltage application facilitates transport of ions through aquifers.

Hexavalent chromium in water is found as the anions chromate, dichromate, or chromic acid, and has been shown to migrate by electrokinetic phenomena towards the anode. At the anode, acidic conditions generated by electrolysis of water and reductive electrochemical conditions promote Cr(VI) reduction to Cr(III), which is a more stable oxidation state with lower solubility and negligible toxicity. Nonetheless, to enhance the remediation process, it is proposed to inject zero-valent iron nanoparticles (nZVI) directly into the subsurface. nZVI have tremendous potential in remediation work due to their high surface area to volume ratio, high reaction rate, and the ability to travel through soil pore space. Furthermore, as compared to macroscale iron particles, nZVI are less expensive and have the added flexibility of being able to inject them into the subsurface at multiple locations and at great depth.

A concern regarding the use of nZVI in natural systems is their tendency to agglomerate and form larger particles, or adsorb to soil particles. This reduces their reactivity and mobility in the subsurface, rendering them potentially inactive. Many conditions affect the degree of agglomeration, such as pH, the presence of multi- or mono-valent cations, and natural organic matter. The likelihood of agglomeration can be determined by the zeta-potential of individual nanoparticles. This is a measure of electrokinetic potential that exists across the interface of all solids and liquids, and as the absolute value of zeta-potential increases, the suspension becomes more stable.

This study explores the use of nZVI in electrokinetic systems for the treatment of Cr(VI) in clay soils. Zeta-potential analysis was performed to determine the stability of surface-stabilized nZVI under experimental conditions. Results indicated a significant negative surface charge on the nZVI, which is increased by the presence of both Cr(VI), and natural organic matter found in all soils and groundwater. The clay soil was also shown to be highly stable under neutral to basic conditions due to a negative surface charge, thereby repelling nZVI in solution.

Integration of nZVI with an electrokinetic soil-column apparatus was conducted to simulate the effectiveness of these two technologies in a natural system. A 20cm column was loaded with a clay soil spiked with Cr(VI), and electric potential was continuously supplied to electrodes housed in reservoirs on either end of the soil column. The reduction and transport of Cr(VI) was investigated with two different methods of nZVI application; as a permeable reactive barrier (PRB) and direct injection. Use of nZVI as PRB effectively reduced Cr(VI) on the cathode-side of the barrier, but had no impact on Cr(VI) on the anode side. Direct injection of nZVI into the electrokinetic cell near the cathode increased reduction rates within the first 72 hours of treatment.

It is found that nZVI can remain stable and dispersed upon injection into a simulated groundwater environment, and move through interstitial space under an electric field to enhance remediation of groundwater and soils contaminated with Cr(VI).

See more of this Session: Water Reuse and Hybrid Water Treatment Processes
See more of this Group/Topical: Topical 1: Water Technology for Developed and Developing Countries (see also Separations Division)

(762e) Starvation of a Mixed Culture of Chlorella Vulgaris and Scenedesmus Sp. for Enhanced Removal of Ammonia From Wastewater Effluent and Its Impact On Lipids Formation

Bala Padmini Lingaraju[1], **Jinsoo Kim**[2], Joo-Youp Lee[1], Kaniz F. Siddiqui[3] and Zhouyang Liu[4], (1) Department of Chemical and Materials Engineering, University of Cincinnati, Cincinnati, OH, (2) Department of Chemical & Materials Engineering, University of Cincinnati, Cincinnati, OH, (3)The Metropolitan Sewer District of Greater Cincinnati, Cincinnati, OH, (4)Environmental Engineering Program, University of Cincinnati, Cincinnati, OH

Starvation of a Mixed Culture of *Chlorella vulgaris* and *Scenedesmus sp*. for Enhanced Removal of Ammonia from Wastewater Effluent and Its Impact on Lipids Formation

Bala P. Lingaraju[1], Jinsoo Kim[1], Joo-Youp Lee[1], Zhouyang Liu[2], Kaniz F. Siddiqui[3]

[1] Chemical Engineering Program, University of Cincinnati, OH 45221-0012

[2] Environmental Engineering Program, University of Cincinnati, OH 45221-0012

[3] The Metropolitan Sewer District of Greater Cincinnati, Cincinnati, OH 45204-2002

Abstract

Microalgae have proven to have the capability to remove nitrogen and phosphorus from wastewater. Investigation of a possible synergy between wastewater treatment and renewable energy production was also considered. However, one of the major barriers against the proposed idea is the slow uptake kinetics. In our preliminary study, a mixed culture of *Chlorella vulgaris and Scenedesmus sp.* starved of nitrogen has shown a faster ammonia uptake rate from secondary wastewater effluent. High density culture and suspended algae has shown faster uptake rate compared to settled algae. In this study, the residual ammonia uptake kinetics has been investigated in a batch system in terms of initial algal seed density using two different seeds grown under nitrogen-deficient and –sufficient conditions in order to determine an initial algal cell density that can achieve 90% ammonia removal within 48 hours. The change in lipids content and profile before and after the ammonia uptake has also been analyzed using solvent extraction followed by gas chromatography. TEM/confocal microscope image analysis and Nile red stain method were also used to validate the change in lipids content.

Keywords: *Chlorella vugaris*, Wastewater treatment, Ammonia removal, Lipids analysis.

See more of this Session: Water Reuse and Hybrid Water Treatment Processes
See more of this Group/Topical: Topical 1: Water Technology for Developed and Developing Countries (see also Separations Division)

(762f) Influent Solids Retained High Biomass Reactor Treating Dairy Manure Wastewater and Kinetics Study

Jingwei Ma, Department of Biological Systems Engineering, Washington State University, Pullman, WA, Liang Yu, Washington State University, Pullman, WA and Shulin Chen, Biological Systems Engineering, Washington State University, Pullman, WA

In large-scale dairies, flushing system is employed as a major means of manure collection for the sake of its efficiency and less labor requirements. The large amount of wastewater needs to be treated for reuse. Anaerobic digestion converts the organic carbon in dairy manure to biogas that can be used as renewable energy to reduce greenhouse and odor gas emission. Flushed dairy manure poses challenges to conventional anaerobic digestion technology. To address this issue, fixed-film digesters with submerged high specific surface area media were introduced to retain active biomass through bacterial attachment. Nevertheless, fixed-film was classified as an inappropriate process for dairy manure anaerobic digestion due to its vulnerability to media clogging. Apparently, a process with no need of exotic media addition would be preferred for anaerobic digestion of flushed dairy manure.

In this study, a strategy using fiber in the dairy manure as biomass carriers for anaerobic digestion of solid wastes was investigated by using anaerobic sequential batch reactor (ASBR) in lab scale, floor scale and pilot scale. Operating parameters, such as settling time, hydraulic retention time (HRT), and organic loading rate (OLR) were evaluated for process optimization. In lab scale experiments, it was revealed that settling time play a key role in biomass retention in form of biofilm. Short settling time favors biofilm formation due to high selection pressure. Bench scale results showed optimum HRT for volumetric methane production rate and methane productivity were 4d and 8 d, respectively. Pilot study confirmed these results and showed 6d HRT should be optimum for ASBR treating flushed dairy manure. A kinetic study was carried out for process control and design. Four basic models, i.e. first-order, Grau, Monod and Chen & Hashimoto were compared with the aim to choose appropriate models. Chen & Hashimoto model based model gave the best simulation with R^2 of 0.99. However, Monod based model showed a poor performance since influent concentration was not include in steady state effluent concentration. At optimized condition (6 d HRT), the volumetric methane production rate of 0.18L/L/d and VS removal of 40% were achieved. It was proved that ASBR using manure fiber as support media not only improved methane production but also reduced the necessary HRT to achieve a similar treating efficiency compared with other study.

See more of this Session: Water Reuse and Hybrid Water Treatment Processes
See more of this Group/Topical: Topical 1: Water Technology for Developed and Developing Countries (see also Separations Division)

AUTHOR INDEX

Adams, Jeffrey Q. 46
Adolfsen, Kristin J. 102
Agasanapura, Basavaraju Nagendra 141
Albert, John 52
Álvarez, Jose R. 128
Antón, F. E. 128
Arena, Jason 100, 146
Armistead, Paul 49
Bai, Fang 119
Balazs, Anna C. 103
Baltus, Ruth E. 141
Banerjee, Kashinath 77
Barber, John 56, 57
Barford, John 120
Bartman, Alex 139
Bhattacharyya, Dibakar 107
Bishop, Bruce A. 104
Brub, Pierre P. 35
Bui, Ngoc 100, 146
Buisson, Hervé 33
Butler, Ethan 148
Campa, Pantaleona 158
Caple, Kevin 79
Cerino-Córdova, F. J. 109
Chan, Charlotte 171
Chapman, Michelle 53
Chellam, Shankar 141
Chen, Shulin 175
Christofides, Panagiotis D. 92
Chung, Tai-Shung 45
Cohen, Yoram 92, 139
Conklin Jr., Alfred R. 131
Cook, Joseph 95
Côté, Pierre 38
Daram, Venkata 108
Das, Tapas 77, 129
Davila-Guzman, N. E. 109
De Pinho, Maria Norberta 121
Desai, Snehal 91
Dhalla, Adil 30
Diaz-Flores, P. E. 109
Duarte, Elizabeth 121
Dubois, Geraud 138
Dutt, Meenakshi 103
Elimelech, Menachem 144
El-Mansour, M. 128
Ferrer-Giné, Joan 139
Freeman, Benny D. 95, 101
Gao, Yiben 145
Garcia, Cristina 158
Ghassemi, Abbas 56, 57, 151
Gil, Ana 121
Giralt, Francesc 139
Giralt, Jaume 139
Gomes, Jewel Andrew 158
González, B. 128
Greenlee, Lauren F. 105, 106
Gu, Han 139
Hashino, Masatoshi 97
He, Jibao 90
Hinds, Bruce J. 32
Hirami, Keisuke 97
Hongo, Teruhisa 118
Hsieh, Wenpin 61
Hsu, Lewis 171
Hua, Chao 119
Huang, C. P. 61
Huang, Liwei 100
Iizuka, Atsushi 118
Iranshahi, Ashkan 137
Ishigami, Toru 97
Jerri, Huda A. 102
Jiang, Peng 94
John, Vijay T. 90
Jovanovic, Goran N. 79
Kamio, Eiji 97
Kanezashi, Masakoto 96
Kasemset, Sirirat 101
Kim, Jinsoo 173
Kim, Young 147
Kim, Yu-Chang 147
Koong, Len Foong 120
Koops, Geert-Henk 34
Kubota, Noboru 97
Kuksenok, Olga 103
Kusanto, Nick 79
Lai, Juin-Yih 42
Lam, Koon Fung 120
Lee, Albert 101
Lee, Chang Hyun 95
Lee, Joo-Youp 173
Lee, Jungho 147
Lee, Kong Hoon 147
Lee, Seockheon 41
Li, Weiyi 145
Li, Yangang 61
Lingaraju, Bala Padmini 173
Lipscomb, Glenn 137
Liu, Zhouyang 173
Luque, Susana 128
Ma, Jingwei 175
Mahmudov, Rovshan 61
Manickam, Seetha 100
Maruyama, Tatsuo 97
Mathews, Alexander P. 80
Matsuyama, Hideto 40, 97
McCullough, Lauren R. 102
McCutcheon, Jeffrey R. 100, 146, 148
McGrath, James E. 95

AUTHOR INDEX

McKay, Gordon 120
McPherson, Gary 90
Mededovic, Selma 130
Meeks, Noah D. 107
Mi, Baoxia 143
Miller, Daniel J. 101
Minhalma, Miguel 121
Moe, Neil E. 56, 57
Moreno, Hector 158
Murad, Sohail 140
Myint, Maung Thein 151
Na, Young-Hye 138
Nagghappan, Lnsp 104
Nason, Jeffrey A. 79
Nordham, David 49
Nunes, Suzana Pereira 39
Oh, Dong-Wook 147
Ohmukai, Yoshikage 97
Panda, Kishora K. 80
Parga, J. R. 158
Pascual, Xavier 139
Peek, Ian 49
Peterson, Eric 158
Pettit, Sandra L. 78
Phillip, William 144
Pirbazari, Massoud 171
Pitera, Jed W. 138
Rahardianto, Anditya 139
Rallo, Robert 139
Rangel-Mendez, J. R. 109
Romero, Alejandro 158
Rufo, Louis J. 78
Sakai, Yuka 118
Sarmento, Georgina 121
Sasaki, Takeshi 118
Semiao, Andrea Jc 126
Sharbat, Ali 56, 57
Sharma, Mukul M. 101
Siddiqui, Kaniz F. 173
Smith, Stanton 104
Speth, Thomas 31
Sunkara, Bhanukiran 90
Tang, Chuyang Y. 145
Thacher, Ryan 171
Torrey, Jessica D. 105, 106
Trussell, R. Shane 37
Tsunashima, Yusuke 118
Tsuru, Toshinori 96
Tung, Kuo-Lun 42
Veite, Jessica 131
Velegol, Darrell 102
Velegol, Stephanie B. 102
Venkatesan, Anand 150
Vivas, B. P. 128
Vora, Ankit 138
Wang, Da-Ming 42
Wang, Jinhui 96
Wang, Xiao-Lin 43
Wang, Yingqing 90
Wankat, Phillip C. 150
Weber Jr., Walter J. 1
Williamson, Kenneth 79
Wilson, Jennifer 131
Wolan, John T. 78
Xie, Wei 95
Xu, Rong 96
Yamasaki, Akihiro 118
Yanagisawa, Yukio 118
Yokochi, Alexandre F. T. 79
Yong, Jui Shan 144
Yoshioka, Tomohisa 96
Yu, Liang 175
Zhan, Jingjing 90
Zhang, Runhu 119
Zhou, Hongde 36
Zhu, Aihua (Richard) 92, 139